Astronomers' Universe

More information about this series at
http://www.springer.com/series/6960

Sujan Sengupta

Worlds Beyond Our Own

The Search for Habitable Planets

Sujan Sengupta
Indian Institute of Astrophysics
Bangalore, India

ISSN 1614-659X ISSN 2197-6651 (electronic)
ISBN 978-3-319-09893-7 ISBN 978-3-319-09894-4 (eBook)
DOI 10.1007/978-3-319-09894-4
Springer Cham Heidelberg New York Dordrecht London

Library of Congress Control Number: 2014953960

© Springer International Publishing Switzerland 2015
This work is subject to copyright. All rights are reserved by the Publisher, whether the whole or part of the material is concerned, specifically the rights of translation, reprinting, reuse of illustrations, recitation, broadcasting, reproduction on microfilms or in any other physical way, and transmission or information storage and retrieval, electronic adaptation, computer software, or by similar or dissimilar methodology now known or hereafter developed. Exempted from this legal reservation are brief excerpts in connection with reviews or scholarly analysis or material supplied specifically for the purpose of being entered and executed on a computer system, for exclusive use by the purchaser of the work. Duplication of this publication or parts thereof is permitted only under the provisions of the Copyright Law of the Publisher's location, in its current version, and permission for use must always be obtained from Springer. Permissions for use may be obtained through RightsLink at the Copyright Clearance Center. Violations are liable to prosecution under the respective Copyright Law.
The use of general descriptive names, registered names, trademarks, service marks, etc. in this publication does not imply, even in the absence of a specific statement, that such names are exempt from the relevant protective laws and regulations and therefore free for general use.
While the advice and information in this book are believed to be true and accurate at the date of publication, neither the authors nor the editors nor the publisher can accept any legal responsibility for any errors or omissions that may be made. The publisher makes no warranty, express or implied, with respect to the material contained herein.

Printed on acid-free paper

Springer is part of Springer Science+Business Media (www.springer.com)

To my daughter Roopsa and to all the kids of her generation
—Let us keep the Earth, our lonely and only home, habitable for the generations to come.

Preface

The whole universe has been a laboratory for the physicists. The motions of the planets in the solar system led Sir Isaac Newton to develop his laws of mechanics. Similarly, the theories of Special and General Relativity are tested in this cosmic laboratory. The physical processes inside a star have enriched our knowledge on nuclear and particle physics. Astrophysics is a branch of applied physics. However, in recent decades, this cosmic laboratory is used for research on chemistry and biology as well. Thus, astrochemistry and astrobiology, two subbranches, have emerged out. Astrochemistry describes the chemical processes in interstellar medium, planets, and in objects that are cooler than a normal star. Astrobiology has emerged out with the discovery of a large number of planets that appear to have an environment appropriate for the origin and evolution of life. Both the physical and chemical processes in a planet offer the possibility of initiating biological processes. Therefore, all the branches of science—physics, chemistry, biology, etc.—are applied to astronomy in order to answer an eternal question of mankind "Is anybody out there?" This exciting development that initiated a new era in astronomy has taken place due to the discovery of a large number of planets outside the solar system. These are the new worlds beyond our own. This has revolutionized our knowledge and understanding about planets, their climates, formation, composition, etc.

I am a professional astrophysicist trained to write highly technical research papers. Unfortunately, current research has become so technical that even another professional astrophysicist cannot understand a research paper fully unless she or he works on the same area of research. Such a technical research paper is reviewed, discussed, and cited by the astrophysicists working on the same area or field. However, when I deliver a lecture in a conference attended by colleagues with expertise on different aspects of astrophysics, I need to make the content quite a bit general. Similarly, when I deliver a lecture to a group of graduate students of science, I need to make the scientific descriptions and explanations even more general. Often the technical aspects are described by introducing analogy with events or objects that are easily understood by the students with basic or advance knowledge on physics and mathematics. However, the task becomes gradually more difficult as the exposure and background of the audience change. From colleagues who work on the same area of research, to the colleagues who work on the same subject but on different area, to scientists working on different subjects, to students with background in physics or astrophysics, to students without specialization on physics, and finally to the students without any science background at all, the degree of difficulties in explaining the most recent developments in scientific research increases severalfold. Therefore, it is the most difficult job, a daunting task to describe and explain the most recent developments in science to laymen.

However, a scientist cannot escape her or his responsibility in reaching to the common people, in sharing the excitement of new discoveries to the common tax payers who contribute in funding most of the research works. On the other hand, the common men are always curious to know the newest discoveries, the newest developments in science and technology. It is no less than a thriller to most of them irrespective of their profession or academic background. As a consequence if scientists do not attempt to reach to the laymen, nonscientists often attempt to bridge the gap by writing popular science book. Once a news reporter, while interviewing me, enquired about my topic of research. I replied "I work on extra-solar planets, also on Brown Dwarfs." The next day it was published "Professor Sengupta works on extra-solar planets also known as Brown Dwarfs"! So scientific documents written by nonscientists may often have a danger of misinterpreting scientific results, even exaggerating the consequence of a new discovery and thus misleading and confusing the common people. This book is my humble effort to fulfill my responsibility of sharing the excitements of the most recent developments and discoveries on astronomy, especially on the planetary science. This is a new experience to me and so it has been a challenging task. On the other hand, by accepting this challenge, I have certainly become more vulnerable to criticism by professional astronomers.

This book is a culmination of several lectures that I delivered to a wide range of audiences—from experts working on my area of research to school students. Often I could not reply or explain several brilliant questions or inquiries by them during or immediately after my lectures. When I found an answer or a possible explanation after several days or months, I was unable to convey it to them. The frustration prompted me to think of writing this book. While writing this book, I imagined that I was delivering a lecture to an audience consisting of three different groups of people: to a group of scientists who were not aware of the current developments in the field of astronomy and in particular of the planetary science; to a group of science students or amateur astronomers who had no exposure to technical aspects and terminologies of advance research; and to a group of laymen who were averse to any unnecessary complexities and did not care about scientific terminologies and mathematical expressions. The first group of people may criticize me for not being able to explain many things properly, the second group may embarrass me by asking several questions beyond my knowledge and understanding and by contradicting my explanations, while the third group may prove that the "Emperor has no cloth." All are welcome.

Preface

This book should be read like a fiction or a story book. But unlike a fiction, all the objects and events described here are real. Also, unlike a story book, each chapter of this book is self-content. Therefore, readers may skip a chapter if it appears to be boring. For this purpose and for reminding the readers some important and related processes, some of the descriptions or explanations are intentionally repeated instead of asking the reader to go back to previous chapters. I hope the experts in this field will pardon me for this intentional repeats. I have put all my efforts to avoid any grammatical errors in the language. However, a few mistakes may still remain. I apologize for that. I hope the scientific contents of the book should dominate the mind of a reader over its literacy value if any.

At the end of the book, a list of websites has been provided. This is no way a comprehensive list but I took help from these websites. These websites may be considered as references for further reading. For the benefit of science students, some simple mathematical relationships and formulae are provided at the end. The main book, however, does not have any reference to these appendices. Therefore, readers without science background should skip these.

Quite a few publishers who invited me to write and publish a book with them strongly suggested to write a textbook and politely refused my proposal for publishing a popular science book for the laymen although astronomy is always a popular subject. I understand that the main reasons are the uncertainty in the marketing of a popular science book and my inexperience in writing any book before, text or popular. Springer has taken that risk and I am thankful to them for publishing the book. I thank Dr. Ramon Khanna, senior astronomy editor of Springer, for several suggestions and guidance.

The amateur astronomers around the world play a vital role in astronomy. Many of the comets, asteroids, and Near Earth Objects are discovered by them. They take active part in planet hunting and in the search for extraterrestrial life (SETI). While professional astronomical organizations such as NASA and European Space Agency (ESA) or European Southern Observatory (ESO) are very generous in providing permission to publish spectacular images of celestial objects, I thought publishing a few images taken by amateur astronomers would not only offer appreciation to their contributions but also might generate interest to common people towards making astronomy as a hobby. I am grateful to Mr. Efrain Morale Rivera of Aguadilla, Puerto Rico, for kindly providing permission to publish some astronomical images from his huge collection.

Ms. Sandra Rajiva, Mrs. Catherina Williams, and Ms. Aarti Dwivedi took the pain to read quite a few drafts of the manuscript and corrected the grammatical errors and suggested several changes in the language. I express my gratitude to them. Apart from the direct help that I received in completing the manuscript, a large number of astronomers motivated and inspired me. Prof. Ronald E. Taam has been a constant inspiration for me. I have learned a lot on various areas of astrophysics from him. I have learned a lot about extra-solar planets and Brown Dwarfs while collaborating with Dr. Mark S. Marley of NASA, AMES. The lengthy, penetrating, and illuminating discussions with Prof. Frank H. Shu during and after my lectures at Academia Sinica, Institute of Astronomy and Astrophysics,

Taipei, have enhanced my knowledge in great extent. I am also thankful to Prof. Vinod K. Gaur for many discussions that enriched my knowledge on the geology of the Earth and Mars. Back in my own country, working in a research institute which has the credit for the first false alarm of extra-solar planets reported some 150 years ago certainly has an advantage. I thank the present director of my Institute, Dr. P. Sreekumar, for his encouragement. Last but not the least I thank my wife Srirupa who took the responsibility of keeping the home in order single-handedly besides managing her career as a teacher.

I shall consider myself successful in this new venture if after reading this book the readers realize by their heart the urgent need of protecting our world, the Earth—an extremely rare planet born and privileged by cosmic coincidence and blessed by a divine process called life.

Bangalore, India   Sujan Sengupta

Contents

1 Hierarchy in the Universe 1
2 A Brief History of the Solar System 15
3 Our Neighborhood: The Solar Family 29
4 Brown Dwarfs: The Missing Link Between Stars and Planets ... 59
5 Discovery of Extra-Solar Planets 67
6 An Amazing Zoo of Planets 83
7 Life: A Delicate Process 103
8 In Search of Another Earth: An Extremely Rare Planet 117
Epilogue 133
Appendix A: Online Resources 143
Appendix B: Some Astronomical and Physical Numbers 145
Appendix C: Extra-solar Planets with the Calculator 147
Index 151

Prologue

Why Care About Other Worlds

"O thievist Night,
Why shouldst thou, but for some felonious end,
In thy dark lantern thus close up the star,
That nature hung in heaven, and filled their lamps
With everlasting oil, to give due light
To the misled and lonely traveller?"
—John Milton
(In *Comus*)

Any normal and developed brain is capable of generating curiosity towards any unknown object or phenomenon. Curiosity is generated by a conscious mind. Once it arises, one or more than one of the five major organs—eyes, ears, nose, tongue, and skin—try to find out an explanation of the object or the phenomenon. If all the organs fail to find out any satisfactory explanation or the primary explanation turns out to have no relevance to the need of the species, the curiosity dies. However, a human brain is much larger than that of any other primate and has much greater functionalities. It uses the information stored in its memory to make a logical sequence and attempts to explain the phenomenon even when the organs fail to analyze it. It is only the human species that can think. But if the information is not sufficient or intentionally not used, then the brain often uses a function—imagination which is probably unique to the human species. Imagination is also a part of thoughts. Unlike curiosity, imagination can be generated by a subconscious brain as well. Imagination gave birth to the concept of demons, gods, and subsequently religions. On the other hand, a combination of the available but insufficient information and imagination gave birth to philosophy.

The fact that the Sun makes the day and the night was realized by humans during the very primitive stage of their evolution. But nothing was known about the Sun and so people of almost all ancient or prehistoric civilizations imagined it as a powerful god. With the advent of agriculture, our ancestors started realizing that the

Sun played a more important role than just giving light and heat to the Earth. Subsequently, they realized the periodicity in the seasonal changes that affected their daily lives. The summer, the winter, and the rainy seasons appear periodically and last for a certain duration of time. So the humans could plan their activities accordingly. Sometimes the summer lasts longer causing damages to crops, sometimes little rain causes severe drought, and sometimes a longer winter causes complete freezing of rivers and lakes. The humans did not have control over such natural calamities. They imagined that the nature was controlled by some superpower, and so they resorted to appease it by prayer. However, the curiosity of the human brain could not be restricted to the surrounding natural phenomena that had direct impact upon their lives but often was extended beyond the objects and events that had no apparent or immediate consequence. In this way, the human beings got curious towards the vast empty space outside their world. They grew interest about the hundreds and thousands of small lights visible in the night sky. Regular observations of those lights provided them a little bit of information. They noticed that the relatively brighter lights moved in a periodic way in the sky while the faint and the twinkling lights remained stationary. With the passage of time, as they gathered more and more information, their understanding headed towards more and more accurate explanations. Thus, quite naturally, the eternal curiosity about other worlds beyond our own was generated. We start asking if there exists in the vast universe another place similar to the Earth, if there is life somewhere else in the universe, if there is intelligent life anywhere in the outer space, etc. As soon as it was established that the Earth was a planet rotating around a star, the Sun, we started believing that such type of planets must be there around the hundreds and thousands of stars in the sky and therefore existence of planets similar to the Earth was very much possible. But where exactly are they located and how are they? Do they harbor life? Although with the rapid advancement of science and technology, our knowledge on the outer space has been enriched enormously, the answer is still elusive. This prompted imaginations and speculations about other worlds, about different kinds of lives, about extraterrestrial intelligence, etc. These speculations or imaginations varied from time to time and from one civilization to other. More than five thousand years ago, the Sumerians imagined amphibian intelligent life with fishy heads but human feet. They were never considered as gods, but the Sumerians speculated that science, arts, and architectures all were taught to the humans by alien intelligent visitors. The Babylonians imagined that the moving lights in the sky were the homes of their gods. According to the ancient Greeks, the Moon was a home of the souls of the dead or the departed people. In fact the Greek philosophers imagined that the Moon had cities with civilized inhabitants. But we even do not know how life originated on planet Earth. The oldest imagination attributes the origin of life on the Earth to a superior being or spiritual force. Most civilizations and religions advocate such idea for the creation of the whole universe including life on the Earth. For example, the Hindus believe that gods and goddesses are superintelligent beings that created the universe, the Earth, and life on it. The speculations and imaginations that often lead to science fictions describe how extraterrestrial intelligence visited the Earth, initiated the origin and evolution of life, and interacted with the human beings in developing various civilizations. The advent of space era has given rise to the imagination of spaceships of extraterrestrial intelligent species visiting the Earth, and thus, Unidentified Flying Objects or UFO becomes a topic of interest among ordinary people. However, such type of speculations lacks any evidence for confirmation and is therefore beyond the scope of science. Nevertheless, the eternal question "Is anybody out there?" remains unanswered.

The whole universe, however vast it is, consists of matter and energy. Even the voids are filled by the relic of the energy that was originated during the very first few minutes after the birth of the universe. The birth event of the universe is popularly known as the Big Bang. Ordinary matter is made of two fundamental types of particles—quarks and leptons. Protons and neutrons are made of a combination of three out of the six kinds of quarks. Protons and neutrons together with electrons that belong to the lepton family make an atom. A few atoms combine to make a molecule. All ordinary matters including the living species are made up of molecules. The dynamic of everything, from the subatomic world to the whole universe, is governed by four fundamental forces—gravitation, electromagnetic, weak, and strong.

In the vast mysterious universe, we see stars, galaxies, nebulas, giant clouds of gas extended over billions of kilometers, planets, comets, asteroids, and several other celestial objects with different sizes, shapes, temperatures, and other properties. All these objects are made of atoms and molecules of various elements and compounds. However, out of all such celestial objects, nature has chosen a planet as the appropriate place for the origin, evolution, and survival of life. In other words, life in the form we recognize it can originate, evolve, and survive in an environment that is available only in a planet. On the other hand, we do not have any clue about life in a form other than what we see in our world. If life exists in a place other than a planet or in a planet that is much different than the Earth, it should have a different form and a different biology completely unknown to us. In fact, there is no way to recognize such a different kind of life if it exists elsewhere. Therefore, it is not possible to investigate or to search for life that is entirely different than that on the Earth. Under such a situation, a systematic and scientific search for life beyond our own world is restricted to the form of life that exists on planet Earth.

The subsequent chapters of this book will present our current understanding and progress towards a rational answer based on scientific analysis of information gathered by using the most advanced technologies available at present. The discovery of 51 Pegasi b, the first confirmed planet outside the solar system, has rekindled the eternal curiosity of mankind to know whether or not worlds similar to our own world exist beyond the solar system. Astronomers have detected more than a thousand of planets outside the solar system, and the number is increasing every month. This is not only taking us steps forward to answer that eternal question but also changing our concept on the nature of planets and planetary systems drastically. Many of the extra-solar planets discovered are entirely different than any of the solar planets. Our knowledge on this field is augmented by the discovery of another kind of celestial objects—"Brown Dwarfs" that very much resemble the giant gaseous planet Jupiter. Brown Dwarfs are considered as the "missing link" between stars and planets. They are lighter than the lightest stars but at least 13 times heavier than Jupiter, although they are as small as Jupiter in size. Astronomers classify these two kinds of objects—Brown Dwarfs and planets—by a common term, substellar mass objects, because the mass of these objects is less than the mass of any star. In fact, the discovery of a large number of giant planets and Brown Dwarfs has revolutionized our understanding of planets to such an extent that the International Astronomical Union (IAU) was compelled to reconsider the definition of planets and to resolve that objects as small as Pluto, even though they orbit around the Sun just like the other large planets, could no more be considered as planets.

However, with the rapid discoveries of other worlds, our knowledge and understanding are changing beyond our imagination. Thus, a global definition of planets could be temporary. Therefore, IAU has limited the definition of planets only to our solar system. A consensus and firm definition of planets outside the solar system is yet to be worked out. In order to address the issue of the existence of life elsewhere, we not only need to find out the conditions under which life can originate, survive, and evolve but also need to address what we mean by life. Therefore, the answer to the question relies upon several factors that provide us a great realization about the uniqueness of our own world. Arthur C. Clarke commented "Two possibilities exist. Either we are alone in the Universe or we are not. Both are equally terrifying." On the other hand, in a vast galaxy containing billions of stars, most of which may have planets, finding another world that is blessed by life is extremely difficult even if the origin of life is a common phenomenon in the universe. That makes us lonely and insignificant. In the words of the anthropologist and naturalist Loren C. Eiseley (in The Immense Journey, 1957), "In a Universe whose size is beyond human imagining, where our world floats like a dust mote in the void of night, men have grown inconceivably lonely. We scan the time scale and the mechanisms of life itself for portents and sighs of the invisible. As the only thinking animals in the entire sidereal universe—the burden of consciousness has grown heavy upon us. We watch the stars, but the signs are uncertain. We uncover the bones of the past and seek for our origins. There is a path there, but it appears to wander. The vagaries of the road may have a meaning, however; it is thus we torture ourselves."

Chapter 1
Hierarchy in the Universe

> *A carbon-based bag of mostly water on a spec of iron-silicate dust around a boring dwarf star in a minor galaxy in an underpopulated local group of galaxies in an unfashionable suburb of a supercluster would look up at the sky and declare, "It was all made so that I could exist!"*
>
> –Peter Walker

Beginning of the Beginning

In the year 1905, Albert Einstein presented his theory of Special Relativity and revolutionized our understanding about matter and energy by telling that they were the same. He derived the famous $E = mc^2$ formula from his realization that space and time have the same status. This realization, in turn, was complimented by his assumption that the speed at which light travels in vacuum is constant in any reference frame. Now what is this reference frame? It is not difficult to visualize the concept of reference frame. Imagine that you are traveling in a train and you throw a ball to a co-passenger in the same direction the train was running. The average speed of the ball should be the distance between you and your co-passenger divided by the time taken by the ball to reach your co-passenger. However, if someone outside the train watches it, one would find that the distance traveled by the ball is the sum of the distance traveled by the train during the time you throw the ball and your co-passenger receives it and the distance traveled by the ball from you to your co-passenger. Consequently, the outside observer would find a higher speed of the ball. The most important thing to notice here is that the time taken by the ball to move from one point to the other will appear to be the same to you and to the outside observer although the distance traveled by the ball would differ. You and your co-passenger measure the speed of the ball inside the train which is one reference frame attached with the train. The outside observer measures the speed of the ball in another reference frame attached to the Earth. One reference frame is

© Springer International Publishing Switzerland 2015
S. Sengupta, *Worlds Beyond Our Own*, Astronomers' Universe,
DOI 10.1007/978-3-319-09894-4_1

moving at a constant speed with respect to another reference frame. If another train passes you with a different speed as compared to the speed of your train and a passenger inside that train measures the speed of the ball, he or she will derive a different speed of the ball. So, the speed of the ball is "relative" to the reference frame wherein it is measured. Einstein considered that light would not follow this rule; the speed of light would remain the same in all reference frames. In order to describe any cosmic event, one has to consider both space and time together. Einstein's Special Theory of Relativity becomes significant only when the speed of an object is comparable to the speed of light. But nothing can exceed the speed of light. Now, if the speed of the ball or the train varies; i.e., if the ball or the train accelerates or retards, Special Theory of Relativity does not apply. For an accelerating reference frame, General Theory of Relativity has to be invoked. The contraction in length and the dilation of time is special relativistic effects whereas the acceleration due to gravitation is a general relativistic effect. Therefore, Special Relativity is a particular case of General Relativity. According to General Theory of Relativity, one can cancel the effect of gravitation locally by moving in an accelerating reference frame such as a free falling lift or aircraft. But it cannot be canceled globally. We know that the universe is expanding. All the galaxies are receding from each other. Since the rate of expansion of the universe changes, the dynamic of the universe is described by General Theory of Relativity.

We do not experience the effect due to Special Relativity in our normal life. However, it becomes significant while describing the dynamics of atomic and subatomic particles inside an extremely hot star or at the center of the galaxies. Similarly, the motion of planets, comets, etc. around a normal star can well be described by Newton's laws. But if an object is extremely compact and much heavier than the Sun, the dynamic of any object near such massive body needs general relativistic treatment. It is, however, worth mentioning that the perihelion precession of Mercury or the bending of light by the Sun is a consequence of General Theory of Relativity which, unlike Newtonian theory, demands that even the path of light is affected by the gravitational force of a massive body.

The universe started by an event or phenomenon popularly known as the Big Bang. By the beginning of the universe, we mean the event or the incident of the origin of space and time. What was before that? There was no "before" because cosmic time started at that point. There was no "what" because space originated at that point. All matters, energy, and everything that we know originated from "Big Bang." We know matter and energy can neither be created nor be destroyed because there is a strict rule of conservation. True, but this conservation rule was violated only once, during the event "Big Bang." It seems everything has come from nothing. The known laws of physics including Einstein's theories of Special and General Relativity break down at this point, and we still don't know what happened during the creation of everything. In fact a fraction of seconds after the creation of the universe, the space was so small that a classical treatment cannot describe it and one has to invoke quantum mechanics to explain the scenario. So, new laws of physics are needed to understand the Big Bang event. However, the amount of matter and energy created at that time remains the same forever.

Along with the matter and energy, four fundamental forces were also originated during Big Bang. Of course at the very beginning they all were unified into one single force and with time they started manifesting in different ways. Recent observation shows that the universe is accelerating which means the rate of expansion of the Universe is increasing instead of decreasing due to gravitational pull. This needs a huge amount of energy that causes repulsion. The acceleration of the universe is either due to the presence of some kind of unknown and bizarre energy known as "dark energy" which is quite different than the energy we encounter in normal life, or we are witnessing something beyond our current knowledge, e.g., something beyond the scope of General Theory of Relativity. If dark energy indeed exists, then the present universe contains 72 % of dark energy, 23 % of dark matter, and only 5 % of visible matter. The gravitational force governs the attraction among matters (even dark matter and dark energy) and plays the most crucial role in determining the dynamics of celestial objects in the universe. The electromagnetic force acts between two electric charges or magnetic poles. Our universe is electrically neutral because the net number of positive charges is exactly equal to the net number of negative charges. However, gravitation does not have such positive and negative quality. Matter and energy always attract each other due to gravitational force. The concept of force is replaced by the curvature or geometry of space time in General Theory of Relativity that describes gravitation. The distance between two points depends on the geometry of the space. In the Euclidean geometry, the shortest distance between two points is a straight line joining the two points. But in non-Euclidean geometry or in a curved space, the shortest distance between two points is not a straight line but is called geodesic. According to General Theory of Relativity, matter–energy determines the surrounding space–time geometry and everything follows that geometrical path in its movement. On the other hand, the space–time geometry tells how much matter and energy are present in a region.

It is not known whether Big Bang created only one universe or several universes. But we know only one universe that is electrically neutral and that has an asymmetry in the sense that there is only attractive force between matters and there is no gravitational repulsion or anti-gravitation. In electromagnetism, there are two kinds of charges or poles—positive and negative charges or North and South Poles. Same charges and same poles repel each other while opposite charges and opposite poles attract each other. There is no such positive or negative matter in the universe we live in. This is known as matter–antimatter asymmetry. Antimatter is like a mirror image of matter and if they come closer to each other, they get annihilated into energy. The universe we live in does not have antimatter. We do not know what caused this matter–antimatter asymmetry. It may be a coincidence that we live in such a universe. We also know that at the very early stage, the universe was dominated by radiation, but at present, it is dominated by matter, or probably by something very weird, called dark energy. The universe cools down as it expands. The signature of this hot early period is carried by the omnipresent cosmic background radiation which was accidentally discovered in 1964 by radio astronomers Arno Penzias and Robert Wilson of AT&T Bell Laboratory at the United States.

The presence of this background radiation with a temperature of 3 K or −270 °C was theoretically predicted by George Gamow and Ralph Alpher. This is known as cosmic microwave background radiation (CMBR) because the wavelength associated with this extremely weak relic energy is about a micrometer or one millionth of a meter. Gamow put the name of his friend, the eminent nuclear physicist, Hans Bethe and the theory got a popular name "Alpha–Beta–Gamma" theory. This theory tells that the elements hydrogen, helium, and lithium were produced during the first 3 min after the "Big Bang" through nuclear fusion of positively charged protons. The stars, galaxies, and the cosmic structure were formed a few million years after the "Big Bang" nucleosynthesis.

In an evolving universe, various events occurred at different cosmic epoch and finally the cosmic structure as we see it today was formed. The duration of the first few epochs was extremely small. We know that light travels about 300,000 km in a second. The time light would take to travel just a centimeter is 1/30,000,000,000 s which is 30 billionth of a second. Six more zeros have to be added after 3 to describe the time taken by light to travel a micron or a micrometer. In order to describe the duration of the early cosmic epochs, one need to add 30–40 zeros after 3 in the denominator. Immediately after the space–time and matter–energy were originated during the Big Bang event, the universe entered in to Planck Era. During this time all the four forces were unified into one and the space–time is speculated to have a foam-like structure. Present laws of physics do not apply to this epoch. The events occurred during this era can be described only if the two major branches of physics—quantum mechanics and General Theory of Relativity—can be unified. In other words, space–time was probably quantized during this period. Planck era ended with the separation of gravitational force from the other three fundamental forces. This is known as Grand Unification era. During this era, the three forces—electromagnetic, weak, and strong forces—did not have separate identity or manifestation. They were unified into one force. Grand Unification era was followed by a brief epoch during which the universe expanded much faster than even light can travel. This epoch is known as Inflationary era. The universe became about ten centimeters in size after this "exponential expansion." Inflationary era should have its imprint on the cosmic microwave background radiation. Hectic search is undergoing at present to detect the predicted signature of this era and to confirm our understanding. After the end of this epoch, strong force that keeps the quarks together got separated from the weak and the electromagnetic forces. This epoch is known as electro-weak epoch. During this epoch, all the particles got their mass by the interaction of a particle called Higgs particle. Higgs particle is discovered very recently in the laboratory. The quark epoch begins after the electro-weak epoch, and during this period, the weak force which is responsible for radioactive decay got separated from the electromagnetic force. During this epoch, quarks, electrons, and neutrinos were formed. The quarks and antiquarks annihilated each other and only the excess quarks survived to form matter. No antiquark survived in the universe. The Hadron era started after the end of quark era and the universe cooled down to a temperature of a few trillion degrees Celsius. Subsequently, the quarks combined with each other to form protons and neutrons and emitted copious amount of a kind of particles called neutrinos. Protons and neutrons belong to the particle family called hadrons while electrons and neutrinos belong to leptons. The Lepton era began when the age of the universe was just 1 s and it lasted for about 10–20 s. This era was dominated by the leptons such as electrons and their antiparticles. The leptons and antileptons annihilated each other to form radiation carried by photons. Therefore, the Lepton era was followed by photon era or radiation era during which the radiation was scattered quite frequently with the protons, neutrons, and the electrons. However, after 3 min of Big Bang, nucleosynthesis began. During this period the temperature of the universe reduced to a billion degrees Celsius and the protons and neutrons combined to form nuclei of hydrogen, helium, and lithium atoms. After about 20 min, the universe cooled down to such an extent that nucleosynthesis halted. The radiation-dominated era, however, continued. After about 250,000 years when the universe cooled down to a temperature of about 3,000 °C and the expansion reduced the density substantially, the free electrons lost their energy and combined with the positively charged nuclei of hydrogen, helium, and lithium. Thus the universe became charge neutral. This period is hence known as recombination era. On the other hand, the photons also became free of any obstacle as they were decoupled from the electrons. Therefore, the universe became transparent. This era is called the era of last scattering or decoupling era. It lasted until the universe became about 300,000 years old. The first stars were formed after 150 million years. The period from 300,000 years to 150 million years after the Big Bang, i.e., the period after the last scatterings took place and before the stars were formed, is known as Dark Age. The universe was dominated by the mysterious dark matter. After 150 million years, the quasi-stellar objects or the quasars were formed out of gravitational collapse of matter. The intense heat from these quasars stripped out the negatively charged electrons from the hydrogen atoms and thus ionized the hydrogen gas. Note that the atoms were without electrons before the recombination era. Therefore, the universe was re-ionized and dominated by charged plasma. This is known as re-ionization era. It lasted for a billion year. The galaxy formation started about 500 million years after the Big Bang. The Sun and the solar system were born when the universe was about nine billion years old.

Supercluster of Galaxies

The visible universe shows structures which have a hierarchy. At very large scale, there are sheet- and wall-like structures made by galaxies and surrounded by large voids that are almost empty of matter. These are called superclusters of galaxies—the largest known gravitationally bound structures in the universe. Superclusters can be extended up to about 300 million light years. The superclusters also contain the so-called dark matter and intra-cluster gas. Our Milky Way Galaxy is a member of such a supercluster known as the Local Supercluster or Virgo Supercluster. The Local Supercluster looks like a pancake. The Virgo Cluster is located near the center of this supercluster and our Milky Way Galaxy is located near the edge of it. The Local Supercluster is about 130–160 million light years long. Similarly there is Coma Supercluster that extends about 300 million light years. Perseus–Pisces at a distance of 230 million light years and Hydra–Centaurus at a distance of about 150 million light years are the two nearby superclusters. The space between the superclusters is filled up by hydrogen gas.

Cluster of Galaxies

The next member in the hierarchy is cluster of galaxies as the galaxies are not isolated but they are grouped in clusters. Clusters of galaxies can be divided into rich and poor. Those having a large number of galaxies are known as rich clusters of galaxies. The rich clusters of galaxies can also be divided into two categories: (1) regular cluster and (2) irregular cluster. Each cluster usually contains 10 % of galaxies, 20 % of inter-cluster gas, and a hooping 70 % of dark matter. The regular clusters are geometrically symmetric with relatively high concentration of galaxies towards the center. The irregular clusters have no geometrical symmetry and there is no center to it. In a regular cluster, the galaxies collide with each other frequently, but the collision between galaxies in irregular cluster is rare. Typically, a regular cluster consists of 1,000 or more galaxies while an irregular cluster consists of 10–1,000 galaxies. The clusters are extended from 3 to 30 million light years. They are separated by an average distance of about 35 million light years. The Milky Way Galaxy is a member of the Local Cluster of galaxies. The Virgo Cluster is a nearby rich and irregular cluster located at a distance of about 50 million light years from us. Virgo Cluster is the dominant member of Virgo Supercluster. Virgo Cluster contains about 250 large galaxies and 2,000 small galaxies. The nearest regular rich cluster, the Coma Cluster, is located at a distance of about 300 million light years and contains about 10,000 galaxies most of which are faint elliptical dwarf galaxies.

Group of Galaxies

The groups of galaxies are placed in the third position of the hierarchy. These are the smallest aggregates of galaxies, and our Milky Way Galaxy belongs to the Local Group of galaxies. Groups of galaxies are usually six to ten million light years long and contain about 20–50 galaxies or less. The Local Group of galaxies contains mostly small dwarf galaxies. Besides the Milky Way Galaxy, two large spiral galaxies—Andromeda and the Triangulum galaxies—also belong to the Local Group. The irregular galaxy Sextans A, another member of the Local Group, is located at a distance of about four million light years from our galaxy. Many of the member galaxies in this group are satellite galaxies of the two most massive galaxies—Milky Way and Andromeda. For example, the Large and the Small

Magellanic Clouds, Ursa Major and Ursa Minor, Leo I, Leo II, Draco, etc. are satellite galaxies of the Milky Way Galaxy while M32, M110, NGC 147, and NGC 185 are some of the satellite galaxies of Andromeda. The Local Group of galaxies is about eight to ten million light years in size.

Galaxies

Galaxies are gravitationally bound system of stars, stellar remnants, black holes, gas, dust, and the invisible dark matter which gravitationally dominates over the visible matters (Fig. 1.1). A galaxy may contain 100 millions to a few trillions of stars depending on its size. Traditionally, normal galaxies are divided into three categories—elliptical galaxies, spiral galaxies, and irregular galaxies. But there are several other types of galaxies such as Seyfert galaxies, quasars, radio galaxies, etc. These are usually known as active galaxies. On the other hand, there are dwarf galaxies and satellite galaxies. The visible universe contains more than 200 billion galaxies.

Based on their visual appearance, Edwin Hubble classified the regular galaxies in a sequence known as the Hubble sequence. The elliptical galaxies are divided into several classes depending on their degree of elliptical shape. They are denoted by E0, E1, E2, E3, etc. E0 type of galaxies are almost spherical, E2 galaxies are more elliptical than the E1 galaxies, and so on. Similarly the regular spiral galaxies are denoted by Sa, Sb, Sc, and Sd depending on how tightly their spiral arms are wounded. The spiral galaxies that have a bar joining the spiral arms through the central bulge are denoted by SBa, SBb, SBc, etc. The galaxies whose shape is intermediate to ellipse and spiral are known as Lenticulars. They are denoted by S0. Lenticular galaxies have no visible spiral arm. Since the arms of the spiral galaxies are contained in an extended disk, these galaxies and the Lenticular galaxies are also referred to as disk galaxies. The Milky Way Galaxy belongs to SBb class of galaxies.

The elliptical and spiral galaxies have a nucleus with a bulge at the center, but irregular galaxies do not have such central region. Most of the regular galaxies except the dwarfs or the satellite galaxies have a supermassive black hole at their center. In active galaxies, the central black hole eats the surrounding matters and releases enormous amount of energy that makes these galaxies much brighter than the normal galaxies.

The Sun is a member of the Milky Way Galaxy which is an ordinary but large spiral galaxy with a nucleus at the center and a nuclear bulge. It is also termed as Galaxy. Like all other spiral galaxies, it has a disk in the form of spiral arms and surrounded by a spherical halo of clouds of stars and star clusters. The visible disk of the Milky Way Galaxy is extended to about 150,000 light years with a thickness of about 2,000–4,000 light years. On the other hand, the halo has a radius of about 300,000 light years. Our Sun is situated at the edge of one of the spiral arms known as the Orion arm. It is about 26,000 light years away from the center of our galaxy

Fig. 1.1 Different types of galaxies (Credit: Efrain Morales Rivera, Jaicoa Observatory, Aguadilla, Puerto Rico; with permission)

(see Fig. 1.2). The Sun rotates around the galactic center at a speed of about 220 km per second, and it takes about 250 million years to orbit the galactic center. The Milky Way Galaxy has two main satellite galaxies—Large Magellanic Cloud and Small Magellanic Cloud. The Milky Way Galaxy possibly has 200–400 billions of stars. For comparison, the neighboring Andromeda galaxy has about a trillion of stars. In fact, it is believed that the Milky Way Galaxy and the Andromeda galaxy are binary galaxies and they are approaching each other. The two galaxies may merge into one elliptical galaxy in four to six billion years.

Stars and Planets

Fig. 1.2 Artist's view of the Milky Way Galaxy with the position of the solar system (Reconstructed by the author from the NASA image)

Stars and Planets

Stars are the basic building blocks of a galaxy. They usually stay in groups—in binaries or in multiple star systems or in star clusters consisting of millions of stars. In a binary system, two stars orbit each other. A multiple star system contains more than two stars. However, our nearest star, the Sun is an isolated star called a field star. On the other hand, the next nearest star, the immediate neighbor of the Sun, belongs to a triple star system. Stars are formed by the collapse of huge and massive molecular clouds. When shock wave produced by a nearby supernovae explosion or collision of galaxies passes through a certain region of the cloud, gravitational instability occurs and the region starts collapsing. This phenomenon triggers instability in the other regions of the same cloud. As a result, various regions of the cloud start collapsing due to their self-gravitation and ultimately a cluster of stars is produced. We note that this process makes all the stars situated in a group because several stars are produced by the fragmentation of the same cloud. Such a group of stars in general is called a star cluster. The group of newly born stars is called open cluster and the group of old stars is called globular cluster (Fig. 1.3). An isolated

Fig. 1.3 Star clusters. The *top image* shows an open cluster consisting of newly born stars while the *bottom image* shows a globular cluster. The globular cluster located at about 37,500 light years away from us is one of the oldest star clusters in the Milky Way Galaxy (Credit: Efrain Morales Rivera, Jaicoa Observatory, Aguadilla, Puerto Rico; with permission)

star such as the Sun is possibly formed out of the collapse of a whole cloud, relatively small and less massive. The gravitational potential energy released during the collapse is converted into heat energy. The temperature at the core of the region gradually becomes so high that thermonuclear reaction starts by fusing two hydrogen atoms yielding into one helium atom. At the initial stage, a star contains a circumstellar disk around it which subsequently evaporates when the star starts shining at its full capacity. At this stage, a less massive star is called T Tauri star and a massive star is called Herbig star. The first star in the universe was born about

Fig. 1.4 The spectral classification of stars. The *lef- hand side* indicates the spectral type and the *right-hand side* shows the corresponding temperature of the photosphere. The *dark vertical lines* are the absorption lines. The photospheric temperature of L and T types of stars or Brown Dwarfs is less than 2,400 K (Illustration by the author)

150 million years after the Big Bang. During that period the universe contained only hydrogen, helium and a little amount of lithium that were synthesized by the Big Bang nucleosynthesis, a nuclear process that took place after the first 3 min of the universe. So the first stars were composed of only hydrogen, helium, and lithium, and they were giant and extremely massive. These first stars lived only for 250–300 million years. But during that period, heavier elements up to iron were synthesized inside them. When they exploded, these elements were spread out to the interstellar cloud. Also during the explosion, the temperature was so high that elements heavier than even iron were produced. The second-generation stars like the Sun were formed out of the collapse of the interstellar clouds which now contain all these heavier elements.

Not all that glitters in a galaxy are stars. A star is defined as an object which (1) balances its inward gravitational pull by the outward thermal pressure, (2) balances the energy loss due to emission by the energy produced inside it, and (3) has a source of continuous energy inside it. The source of this continuous energy is the thermonuclear process known to most of the common people as hydrogen bomb. When hydrogen atoms get depleted, helium starts burning. Helium burning produces so much heat that the outward thermal pressure exceeds the gravitational pull and hence the star starts expanding to a state called red giant. When energy production completely stops, gravitational force becomes dominant and the star starts contracting and ultimately becomes either a white dwarf or a neutron star or a black hole depending on its mass.

Astronomers deduce the properties of a star by passing its light through an instrument called spectrograph. This instrument splits the normal starlight into different colors associated with different wavelengths and provides the spectrum. The light emitted at different wavelengths gets absorbed by the elements present in the photosphere of the star. This absorption is seen as dark lines in the spectrum of a star (see Fig. 1.4). The absorption features varies from star to star depending on the temperature of their photosphere and the elements present in the photosphere. Depending on their spectra, stars are classified into nine main types—O, B, A, F, G, K, M, L, and T. The O type stars are the hottest. There is one more class recently added in the spectral classification, Y. Y type of stars are the coolest. Some of the L-type stars and all the T- and Y-type stars are really not stars. They are called Brown Dwarfs because they are so cool that they could not sustain nuclear burning in their core. But they are formed in a similar way a star is formed. We shall discuss this kind of objects in Chap. 4. Depending on the temperature, different types of stars look different to us. O-type stars are blue, B-type stars are white-blue, A-type stars are white, F-type stars are yellow-white, G-type stars are yellow, K-type stars are yellow-orange, and M-type stars are orange-red while L- and T-type stars are red-brown. Y-type stars or the Y Brown Dwarfs are so faint that they may not be visible as normal light source. The infrared light from them is detected. Each spectral class is further divided into ten subclasses, from 0 to 9, according to the Harvard classification scheme. For example, G-type stars are subdivided into G0, G1, G2, G3, G4, G5, G6, G7, G8, and G9. Among all the G-type stars, G0 stars are the hottest and G9 stars are the coolest. So it is for other classes. Each spectral type is also divided according to the atmospheric extension which is determined by the temperature of the star. This is known as the Morgan Keenan classification. Accordingly, there are six types of stars represented by roman digits. Class I stars are supergiants, class II stars are bright giants, class III stars are giants, class IV stars are subgiants, class V stars are dwarfs, and class VI stars are subdwarfs. According to both Harvard and Morgan Keenan classifications, the Sun is a G2V-type star with photospheric temperature about 5,500 °C. Therefore, Sun is a G-type yellow dwarf star. Photosphere of a star is the outer region of the star from which light comes to us. The region inside the photosphere is much hotter and the center of the Sun has a temperature as high as a few millions of degree Celsius. The spectra that we observe are originated from the photosphere only. Below the relatively transparent atmosphere or photosphere, there is a region where matter boils, moves upwards and downwards just like the boiling water. This region is called the convective region. Convective region is opaque. The core of a star lies below this convective region. It is the core that produces the energy through nuclear process. Therefore, the power generated by a star actually depends on the mass of this core. The more massive the core, the brighter the star is. On the other hand, brighter stars have shorter life spans because they exhaust all the fuel in shorter period. The estimated age of the Sun is about four and a half billion years. It will last for another four billion years before all the hydrogen fuel is depleted inside the core.

Not all stars have planets around it. Our present understanding on the formation of planets (see Chap. 2) implies that only those stars that have high amount of heavier elements (known as metal rich stars) can form planets. Also it is very unlikely that a binary or a multiple star system can have planets around any of the stars in the system. Under such situation dynamical stability of the planetary system is usually not possible. So, either the planet jumps into one of the stars or ejects out from the system. Nevertheless, planets around a few binary stars have been discovered recently. We shall discuss all these points in the subsequent chapters. It is, therefore, quite apparent now that stars are placed at the bottom of the hierarchy in the universe while planets are just companions to the stars. Comets, asteroids, etc. also accompany the stars.

If we consider from the smallest to the largest, we find that the typical distance between two stars in the galaxy is about 4 light years; for a large galaxy such as Milky Way Galaxy, the distance from one end to the other is about 300,000 light years; the typical distance between two galaxies in a group is about 3.5 million light years while that between two clusters of galaxies is about 40 million light years. Finally, the time taken by light to travel from one edge to the other of a supercluster could be about 300 million years. The whole visible universe is as large as a few billion light years and it is expanding.

If an extraterrestrial intelligent friend at the furthest corner of the universe would like to send you a letter through an imaginary postal mail, according to the hierarchy in the universe, it should be addressed as:

To

Your name

1. City or town
2. District
3. State
4. Country
5. Continent (optional)
6. Earth (the 3rd planet)
7. The solar system
8. The Milky Way Galaxy (Orion spiral arm)
9. The Local Group of galaxies
10. The Local Cluster of galaxies
11. The Virgo Supercluster

The first four items are due to political hierarchy, the fifth item is due to geographical hierarchy, and the remaining five items are due to celestial hierarchy.

We shall end this chapter by noting that the stars, even in our nearest galaxy, the Andromeda galaxy, are barely resolved visually due to the enormous distance between the two galaxies. Therefore, at present, there is no way to know the worlds beyond our Milky Way Galaxy.

Chapter 2
A Brief History of the Solar System

> *This world was once a fluid haze of light,*
> *Till toward the centre set the starry tides,*
> *And eddied into suns, that wheeling cast*
> *The planets: then the monster, then the man.*
>
> –Alfred Tennyson
> (*In* The Princess, 1847)

The Solar System: Our Neighborhood

From the year 1930 to the year 2006, nine objects in the solar system were considered as planets. The planets Mercury, Venus, Earth, and Mars are known as terrestrial or rocky planets because they have a solid surface like that of the Earth. The planets Jupiter, Saturn, Uranus, and Neptune are known as Jovian planets. They all are giant gaseous objects. Not much was known about Pluto. Even the mass of this planet was not known until the end of the last century. Now we know that Pluto is a rocky object and it is much smaller than even our Moon. The terrestrial planets are also known as the inner planets and the Jovian planets are known as the outer planets because an asteroid belt between Mars and Jupiter separates them. This asteroid belt has many small objects orbiting around the Sun and some of them are as large as Pluto. All the planets except Pluto are orbiting in the same plane. We also know that the planets are not the only members of the solar system. There are a large number of satellites, comets, asteroids, and planetesimals (small, rocky celestial objects formed during the birth of the solar system). In 2006, the general assembly of the International Astronomical Union, the "Parliament" or "Senate" of astronomers, decided a definition of the planets and resolved that Pluto should no more be considered as a planet. So, if we go by that definition, then the number of planets reduces to eight. Pluto and other objects similar to the size or mass of Pluto are now considered as Dwarf Planets. We shall discuss it later on.

If one is born and brought up in a small town, has never traveled outside it, and doesn't know much about the people outside the town, then the small town becomes one's very own world irrespective of one's knowledge that there exist other towns and cities and even countries. So, in this book we consider the solar system as "our world" because we know the parent of this planetary system, the Sun, and the neighbors of the Earth the best. All the planets and the Sun have mutual influence on each other gravitationally. Also the formation and evolution of the Earth are coupled with the formation and evolution of all the other solar objects. The chemical composition or the elemental abundance of all the solar system objects is also the same because they all were born out of the same molecular cloud called the solar nebula. Even the fate of all the planets in the solar system is almost the same as that of the Earth.

Our present understanding based on several observations also implies how unique the solar system is, at least within our galaxy. However, before we discuss about the individual planets and their discoveries, let us see how a planetary system in general or the solar system in particular was formed.

How the Solar System Was Formed

It is believed that all the solar system objects—the Sun, the planets, the Moon, asteroids, comets, etc.—were formed at the same time and out of the same nebula or interstellar cloud. Therefore, the solar planets and their parent star, the Sun, have almost the same age—about 4.6 billion years. Now, the question is how a star similar to the Sun is usually formed. A star is formed by the gravitational collapse of a huge cloud of molecules and dust in the interstellar medium. The cloud is so large that light takes about 3–4 years or more to travel from one end to the other. It is spherical in shape and is spinning slowly around its own axis of rotation. The outward pressure caused by the heat of the cloud exactly balances the gravitational pull towards its center. This is called hydrostatic equilibrium or hydrostatic balance. Now, imagine that a slight disturbance is given by compressing a certain part of the cloud. This would cause a perturbation in the cloud and the disturbance will travel through the cloud at the speed of sound. This is analogous to the propagation of the disturbance that we observe when we drop a stone in the stagnant water of a pond or a tank. The speed of sound in any medium is directly proportional to the square root of the temperature of that medium. Therefore, the disturbance travels faster if the medium is hotter and it travels slower if the medium is cooler. Suppose this disturbance takes a time t to travel a distance R within the cloud. As this disturbance passes through the cloud, the pressure tends to bring back the medium to its original stable condition. At the same time the gravitational force tries to pull matter towards the center so that there is a tug of war between the outward pressure and the gravitational force in this disturbed or perturbed region. The matter is displaced at a speed called the free fall velocity. What is it? If you throw a stone upwards, it goes up with a certain speed that keeps on decreasing with the height, and at some point, it comes to the rest and then falls back to the Earth. The motion of the falling stone is directed towards the center of the Earth. The speed at which the stone falls depends on the mass and the radius of the Earth and not on the size or mass of the stone. That is why objects of all kinds, irrespective of their mass or size, fall onto the Earth with the same speed. This phenomenon was demonstrated first by Galileo Galilei. The velocity at which any small object, initially at rest, falls towards a massive object under the influence of its gravitational attraction is called the free fall velocity. So, the adjacent matter within the disturbed region of the cloud would move at the free fall velocity towards the center. Now, if the time taken by the disturbance to travel the distance R is shorter than the time taken by the matter to make a free fall of distance R, then the pressure wins the race and the disturbed region is restored to its initial condition. This means nothing happens to the cloud by the disturbance.

But, if the region is sufficiently cool so that the disturbance does not propagate fast enough and if R, the size of the region, is sufficiently large, then the time taken by the disturbance to travel across the region becomes longer than the free fall time. In this situation, the gravitational force wins the race and the whole region starts collapsing. The minimum size of a region or the critical length of a region at which this is possible is called the Jeans Length after the name of Sir James Jean who pointed out this during 1940. The phenomenon is called the Jeans instability. The total mass contained in a spherical region of diameter equal to the Jeans Length is called the Jeans Mass. Therefore if the mass of a region is equal to or more than the Jeans Mass, then any external disturbance exerted to this region would initiate gravitational collapse. Of course the Jeans Length depends on the initial temperature of the cloud. Further investigations on Jean's analysis show that if one region of a large cloud starts collapsing, the other regions surrounding it also get disturbed and as a result the whole cloud gets fragmented into several collapsing regions. This triggers the formation of stars with different masses depending on the mass of the different collapsing regions.

Now let us consider one such collapsing spherical cloud which is also spinning around its own axis. As the size of the sphere reduces, the spin velocity increases because the angular momentum of the cloud has to be conserved. Angular momentum of a rotating body is the product of its rotational speed and the moment of inertia. The moment of inertia is determined by the distribution of matter inside the body. For a spherical body, it is proportional to the square of the radius. Hence, if the radius decreases, the moment of inertia also decreases. Therefore, to keep the product—the angular momentum unaltered—the rotational speed increases. Now, the centrifugal force due to the spin acts in the opposite direction to the gravitational attraction and produces an equatorial bulge. As a result, the increase in the rotational speed and hence the increase in the outward centrifugal force slow down the inward flow of matter in the equatorial plane. As the collapse continues, the rotational speed keeps on increasing. At some point the rotational speed of matter in the equatorial plane becomes so high that the outward centrifugal force acting on the matter exactly balances the inward gravitational pull. This is known as the breaking point of the rotation. If the rotational speed exceeds this breaking point, then matter overcomes the gravitational force and starts escaping from the equator. Therefore, at the breaking point of rotation when the centrifugal force is balanced by the gravitational pull, the matter in the equatorial plane only rotates around the central high-density region called the protostar and does not move towards the center of the system. All the remaining matter falls onto the equatorial plane. Consequently the shape of the cloud becomes ellipsoidal and finally becomes a thick disk rotating around the protostar. This is known as a protostellar disk. At the same time, the central collapsing region, i.e., the protostar which remains spherical, becomes denser and hotter due to the release of gravitational potential energy. If the mass of this protostar is greater than about 8 % of the present mass of the Sun, then it becomes so hot that two hydrogen atoms are fused releasing enormous amount of nuclear energy. If the process of nuclear burning at the core of the region continues, then the protostar becomes a main-sequence star. However, if the mass of the central region is less than the critical mass for nuclear burning, then the heat generated by the gravitational potential energy becomes insufficient to sustain the nuclear process. In that case a Brown Dwarf is produced instead of a star. We shall discuss this situation in Chap. 4.

Is there any upper limit on the mass of the protostar? The answer is yes. If the protostar is about a hundred times more massive than the Sun, the star becomes unstable and cannot survive even a million years. This kind of stars is called Wolf–Rayet stars after their discoverers Charles Wolf and Georges Rayet. Wolf–Rayet stars are at least 20 times heavier than the Sun and are very unstable. Therefore, in order to form a main-sequence star, the mass of the protostar should lie between 0.08 and 100 times the mass of the Sun. Of course this range is slightly different for stars with different chemical composition or elemental abundance. For example, a protostar may be poor in heavy elements such as carbon, oxygen, nitrogen, etc. and composed mainly of hydrogen and helium. Such a protostar needs to be much heavier than 0.08 times the mass of the Sun. Otherwise it cannot become a normal or main-sequence star. All the normal stars are born through the process described above. The disturbance or the perturbation that triggers the fragmentation and collapse of the cloud may originate from the shock wave produced by the explosion of a nearby star (known as supernova) or from the collision between two galaxies or by any other energetic transient phenomenon.

Now, the evolution of the surrounding disk varies depending on the material of the disk. It also depends on the mass and hence the temperature of the protostar and subsequently on the brightness of the newly born star. If the initial star forming cloud is composed only of primordial light elements—hydrogen, helium, and lithium—most of the matter would be used in the formation of the central star and the remaining matter would blow away or disperse in a short period of time. Therefore, the first-generation stars that were born out of the primordial light elements or the stars which lack sufficient amount of heavy elements (metal-poor stars) cannot have planets around them as there was no residual material available around them to form planets.

On the other hand, if a star forming cloud is composed of heavy elements that were synthesized inside the first generation of stars and during the explosion of the first generation stars, the protostellar disk may survive even after the star is born. The disk is then called a proto-planetary disk because all the planets, planetesimals, comets, asteroids, etc. are formed out of the material within this disk. However, if the mass of the newly born star is a few times the mass of the Sun, the strong radiation energy emitted by the star blows off the disk. Therefore, formation of planets is not possible around massive and very bright stars even if they are rich in heavy elements or metal rich. Note that the astronomers call all elements heavier than hydrogen and helium as metals. So, if the newly born star is like our Sun or is fainter than it or even a Brown Dwarf that has failed to become a star, the proto-planetary disk formed around it would initiate the planet formation process. At first, the dust particles of size as small as a few centimeters stick to each other by electrostatic force. Once they become sufficiently large, they attract matter through their gravity and several planetesimals of size about a kilometer are produced in this process. These planetesimals collide with each other to form bigger objects that are the embryos of planets.

The next stage of the evolution is governed by the mass of the embryos and their proximity to the stars. The embryos that are very near to the star would get strong illumination over their surface and so become very hot. As a result they lose most of their gaseous components composed of light elements such as hydrogen and helium. This gives rise to rocky planets such as Mercury, Venus, Earth, and Mars. On the other hand, the outer bodies are cooler and so they would retain the gaseous components. These outer bodies are more massive as well and hence their gravitational attraction is also very strong. Therefore, matter cannot escape easily. This gives rise to the Jovian planets such as Jupiter, Saturn, Uranus, and Neptune. Note that the entire planet is made of gas and there is no solid rock inside any Jovian planet. Of course, the pressure near the center of such planets should be so high that the gas becomes liquid at even very high temperature.

The actual mechanism of planet formation involves several stages and many complicated dynamical and radiation processes. Astronomers are still working on it by using sophisticated computer codes. However, the above description provides a basic idea on the formation of any planetary system in general and the formation of the solar system in particular. Interestingly, the other planetary systems discovered so far are all drastically different from the solar system. Contrary to our current understanding, they all have giant gaseous planets very near to their parent stars. This might be caused by the frictional force in the proto-planetary disk and astronomers call the mechanism as "planetary migration." Under this hypothesis, the giant planets were born far away from the star but then migrated very near to it.

How Many Planets? The History of Controversy

Aristotle's World: Seven Planets for Two Thousand Years

Now we know how a large and massive interstellar cloud gives birth to different types of objects whatever be their names. The name is essential to identify a group of objects with similar nature or physical properties. Out of each collapsing cloud, one central object is born which produces its own energy through nuclear burning. We call it a star. So if we identify an object as a star, we understand that (1) it has its own source of continuous energy, (2) the amount of energy it releases is balanced by the amount of energy it produces, and (3) the outward radiation force is exactly balanced by the inward gravitational force. This is the definition of a normal star.

The word "planet" is derived from the Greek word "planetos" means wanderer. Thousands of years ago people of various ancient civilizations noticed that hundreds and thousands of lights moved around the night sky. The Babylonians, the Greeks, the Mayans, the Egyptians, and the Indians noticed that out of the hundreds of wandering lights, seven were very bright and distinct. They knew that human life on the Earth was affected by the two bright light sources, the Sun and the Moon. These two sources were the brightest among all the other light sources in the sky. But they also noticed that five other bright lights, although did not have any impact on life, behaved similar to the Sun and to the Moon in their motion and in their appearance.

It is the famous Greek scholar and philosopher Aristotle (384–322 BC) who for the first time integrated physics and astronomy by postulating the concept of aether. Aristotle applied arithmetic and mainly geometry to describe the dynamical behavior of these bright lights. According to him, all these bright lights were made of something called aether and they all, including the Sun and the Moon, were rotating around the Earth. The Earth was, however, made of water, fire, air, etc. Aristotle assumed that nature preferred absolute symmetry or perfection and so the orbits of all these seven objects were perfectly circular (Fig. 2.1).

So, Aristotle's world was geocentric. A little after Aristotle, another Greek scholar, Aristarchus (310–230 BC) measured the sizes of the Sun and the Moon and their distances from the Earth in term of the Earth's radius. He derived that the Sun was actually six to seven times larger than the Earth. Considering that smaller objects should rotate around much bigger objects, Aristarchus concluded that the Earth should rotate around the Sun. However, due to the influence of Aristotle's philosophy, there was no taker of his finding. But this was the first attempt to understand that the world is actually heliocentric and not geocentric.

During the second century AD, Klaudios Ptolemaios (85–165 AD), best known as Ptolemy, provided a detailed and complicated mathematical version of Aristotle's geometrical description of the geocentric world. Ptolemy mathematically reproduced the orbits of all the seven celestial "wanderers," their distances from the Earth and the speed at which each of them rotates around the Earth. According to Ptolemy, there were seven planets, and according to their distances from the

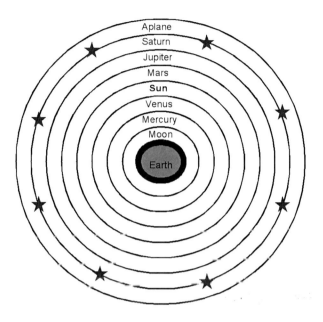

Fig. 2.1 Aristotle's geocentric world. It was visualized as a combination of 55 crystal spheres containing various celestial objects. The sphere after Saturn contains the distant stars (Illustration by the author)

Earth, as he calculated, they were Moon, Venus, Sun, Mars, Jupiter, and Saturn. The names of all these seven wanderers were given after the Greek gods and goddesses. But the subsequent emergence of a strong empire, the Romans, changed the names after their gods and goddesses. The present names of all the planets were derived from the Roman mythology.

During the time of Ptolemy, two strong religions were born, Christianity and Islam. The Roman Catholic Church adopted Ptolemy's concept of a geocentric world and dictated the acceptance of it so widely and strongly that the Ptolemy's view of the world existed for more than a thousand years. Note that the Earth was still not considered as a planet and it carried a special status in the universe.

Beginning of Modern Astronomy: Six Planets for a Few Decades

Modern astronomy began its journey during the early sixteenth century. Around the year 1510 AD—nearly 1,800 years after Aristarchus realized that the Earth was not the center of the world—Nicolas Copernicus (1473–1543), a polish astronomer, came out with the revolutionary proposal that the Sun was the center of the world and the Earth revolved around it. In fact Copernicus went a step forward. According to Aristotle's opinion, all objects except the Earth were made of aether and they moved in a perfectly circular orbit around the Earth which was made of air, fire, water, and earth. Copernicus proposed that although the Earth was not made of aether, it moved around the Sun and the Sun did not move although it was made of

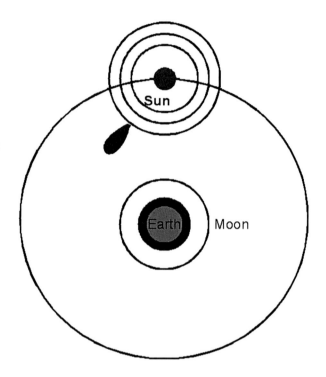

Fig. 2.2 Tychonic system of the world. This is partly heliocentric, partly geocentric. The orbits of three planets are shown centered on the Sun. The orbits of the Moon and the Sun are centered on the Earth. The comet appeared in 1577 AD is shown orbiting the Sun (Illustration by the author)

aether. Copernicus proposed that the Sun was the center of motion of the planets and the Earth was the center of motion of the Moon. So, for the first time, the Earth got the status of a planet and the Sun was no more a planet whereas the status of the Moon was undefined. However, the Roman Catholic Church did not accept the Copernican world. No astronomer, except the Italian thinker Giordano Bruno (1548–1600), supported the Copernican concept of a heliocentric world. During the latter half of the sixteenth century, the Danish astronomer Tycho Brahe (1546–1601) gathered a good amount of information through observations. Tycho Brahe proposed a scenario which was sort of a compromise between Aristotle's world and Copernicus' world. The "Tychonic" world was partly heliocentric and partly geocentric.

In this world, all the planets—Mercury, Venus, Mars, Jupiter, and Saturn—move around the Sun forming one system. The Moon and the Sun along with this system move around the Earth (see Fig. 2.2). Tycho Brahe also discovered a comet and realized that not all heavenly bodies moved in perfectly circular orbits—a contradiction to Aristotle's concept. However, Tycho Brahe in his proposal retained Aristotle's concept of a geocentric world.

So, the number of planets remained the same.

The Kepler's Laws

Finally, the German mathematician Johannes Kepler (1571–1630) who used the observational records of Tycho Brahe and the Italian physicist Galileo Galilei (1564–1642) established the fact that the Earth indeed moves around the Sun. Galileo used the newly invented telescope to observe the sky. This type of telescope uses a convex lens and is known as Galilean telescope. In order to acquire the observational records of Tycho Brahe, Kepler joined Tycho as his assistant. After Tycho passed away, Kepler got access to a large amount of data. By using it, Kepler discovered the three laws of the planetary motion. The first law states that the planets orbit the Sun in elliptical paths—a sharp contradiction with the Aristotelian thought of perfect circular and the geocentric world. The second law states that the straight line joining the Sun and a planet sweeps out equal areas in equal time. The third law states that the square of the period of revolution of any planet is directly proportional to the cube of the semimajor axis. The semimajor axis of an ellipse is the longest distance from the center of the ellipse to its edge. The shortest distance is called semiminor axis. Together, these three laws are known as Kepler's laws. It was Sir Isaac Newton (1642–1727) who explained these laws by his theory of gravitation.

So how many planets are there? We can summarize it. (1) During the ancient civilizations prior to the Greek and the Babylonian civilizations, seven wandering stars—Mercury, Venus, Moon, Sun, Mars, Jupiter and Saturn—were known. (2) For the next 2,000 years, from the Aristotle's era to the middle of the sixteenth century, these seven objects were termed as planets with the Earth as the center of the motion of all the seven planets. The Earth was also thought to be physically quite different from the planets. (3) In the sixteenth century, it was realized that the Earth was also a planet orbiting around the Sun. Kepler demonstrated that there were six planets—Mercury, Venus, Earth, Mars, Jupiter, and Saturn. So the Sun is no more a planet, neither the Moon is. However, the status of the Moon was undefined although it was known that the Moon revolved around the Earth.

Discovery of Jupiter's and Saturn's Moons: 16 Planets

Galileo Galilei, one of the fathers of modern science, was born 21 years after the death of Copernicus. Galileo had regular communication with Johannes Kepler and he too realized the correctness of Copernicus' concept of a heliocentric world. While Kepler was interested in the mathematical formulation of the dynamics of the planets, Galileo devoted most of his life in experimenting on dynamics and optics. During the early seventeenth century, when Kepler had just published his laws of motion for the planets, Galileo constructed his telescope that made any object larger by 20 times. At first this instrument was used by the merchants to track their ships and also by the military. But at some point of time, Galileo started using it to watch the night sky. The first thing that he observed was the Moon. He immediately realized that the Moon was like the Earth and not made by aether as thought by Aristotle. He discovered the craters and mountains of the Moon. He also found the sunspots and realized that the Sun too was not perfectly spherical. Subsequently, Galileo made his most important discovery, four small objects rotating around Jupiter. Galileo called them the Medicean stars. This discovery made him to understand that the Earth was not the only center of motion as advocated by Aristotle but Jupiter was also a center of motion for some objects. Now, if both the Earth and Jupiter could be the center of motion, then why not the Sun? So, Galileo immediately realized that Copernicus was correct. However, the Catholic Church opposed the concept and forced Galileo to withdraw his support for a heliocentric world. This is now a history and we will not go into that.

Now, what happened to Galileo's four "Medicean stars"? Although Galileo retracted his view under the pressure of the Church, many other people observed these four objects. They were soon considered to be planets. If so, then the Moon should also be a planet. In this way, by the year 1610, we had a total of 11 planets—6 orbiting around the Sun as proposed by Copernicus, 1 orbiting around the Earth, and 4 orbiting around Jupiter. All of them follow Kepler's laws. Of course Kepler's laws were slightly modified by replacing the Earth as the center of motion for the Moon and Jupiter as the center of motion for the Galilean planets. Most probably Kepler realized that the six planets orbiting around the Sun should be different from the five objects orbiting around the Earth and Jupiter. In 1655, about half a century after the discovery of Jupiter's moons, Christiaan Huygens discovered a moon of Saturn called Titan. After 4 years, Huygens reported the presence of the ring around Saturn. During 1671 and 1684, Jean-Dominique Cassini, an astronomer at the Observatory of Paris, discovered four more moons orbiting around Saturn.

So, by the end of the seventeenth century, the number of planets became 16—6 orbiting around the Sun, 1 orbiting around the Earth, 4 orbiting around Jupiter, and 5 orbiting around Saturn. Note that the Sun was no more considered as a planet but the Earth was a planet and all the satellites discovered were also considered as planets. The whole scenario changed within two centuries—from 7 to 6 and then to 11 and finally to 16 planets. If we don't care about the name or the status of the planets and their satellites, then the most important development during these two centuries was the understanding that (1) the Sun is stationary and the planets rotate around the Sun in elliptical orbits, (2) the Earth is also a planet, and (3) the discovery of ten more objects in the solar system—Tycho's Comet, four objects around Jupiter, and five objects around Saturn.

Discovery of Planets Beyond Saturn: Number Changes

Sixteen was the highest number ever considered for the planets. However, Huygens observed that the Moon and the objects orbiting around Jupiter and Saturn were quite different from the six planets revolving the Sun. So he termed the ten objects rotating around the Earth, Jupiter, and Saturn as secondary planets or moons or satellites. The six objects rotating around the Sun were termed as primary planets. In the mean time some interesting phenomenon was observed. Edmond Halley (1656–1742) observed a comet in 1682 and calculated its orbit by using the laws discovered by Sir Isaac Newton. Halley's calculation was published in 1705. Clearly this was not a planet. In fact Tycho's Comet was also not considered as a planet. In 1758, a comet was spotted and it was identified as the same comet that Halley observed and predicted for its reappearance. It revealed the fact that Halley's Comet orbited the Sun and followed Kepler's laws but it was not a planet. So, not everything that orbits the Sun is a planet. By the second half of the eighteenth century, it became usual to call the objects orbiting around the Earth, Jupiter, and Saturn as moons and consequently the number of planets in the solar system reduced back to six. The ten satellites and all the comets were of course considered as the member of the solar system. Sir Isaac Newton discovered his laws of gravitation as an interpretation of Kepler's laws. In fact it is known that Newton wrote his famous book "Principia" on constant pursuit of Halley. Astronomy plays such an important role in science! So, in less than 300 years, the number of planets changed from 6 to 11, to 16, and back to 6.

In 1766, Johann Daniel Titius discovered a relationship among the distances of various planets from the Sun. The distance between the Earth and the Sun is called one Astronomical Unit or AU. One AU is equal to 149.6 million km. Titius found that the distances of the planets followed an arithmetical pattern. From the Sun, the distances to the planets can be expressed as $0.4 + (0 \times 0.3) = 0.4$ AU, $0.4 + (1 \times 0.3) = 0.7$ AU, $0.4 + (2 \times 0.3) = 1$ AU, $0.4 + (4 \times 0.3) = 1.6$ AU, $0.4 + (8 \times 0.3) = 2.8$ AU, $0.4 + (16 \times 0.3) = 5.2$ AU, and $0.4 + (32 \times 0.3) = 10$ AU. This means the distance to Mercury from the Sun is 0.4 AU, the distance to Venus from the Sun is 0.7 AU, and so on. Note that the first and the last numbers remain unchanged while the middle one changes in a specific way (0 for Mercury and power of 2 for other planets). However, it was found that the planet Mars was at a distance of 1.6 AU, but Jupiter was at a distance of 5.2 AU from the Sun. There was no planet at a distance of 2.8 AU from the Sun. Six years after Titius published his formula, Johann Elert Bode realized that there should be another planet at a distance of 2.8 AU from the Sun, i.e., between Mars and Jupiter.

So the Titius–Bode formula predicted a planet between Mars and Jupiter. Extrapolation of the formula beyond Saturn also predicts the presence of a planet at a distance of $0.4 + (64 \times 0.3) = 19.6$ AU. In 1781, William Herschel discovered a planet which was at a distance of 19.18 AU from the Sun, very closed to the predicted distance. Historically, this object was observed first by John Flamsteed in 1690 but he could not realize that it was a planet. Bode suggested the name of the planet as Uranus. So the number of planets was back to seven, as it was before the sixteenth century.

By the beginning of the nineteenth century, astronomers had two powerful theoretical tools, one was the Titius–Bode empirical formula that gives the distance of the solar planets and another was the Kepler's law which gives the orbital motion of a planet. With these two tools, astronomers could predict the presence of new planets. During the early nineteenth century, it was obvious that the planet Uranus was not exactly following Kepler's law. It was also found that Halley's Comet was delayed in appearing to its predetermined position. All these observations made people to hypothesize another planet, planet X whose gravitational field was considered to be responsible for the deviation in the motions of Uranus and Halley's Comet. Finally, in 1846, after more than half a century of the discovery of Uranus, the hypothetical planet X was discovered and was named as Neptune. Neptune remained the farthest planet in the solar system for almost a century until the discovery of Pluto in 1930.

Eleven Planets for Four Decades

Was Neptune the eighth planet in the solar system? No, Neptune was the thirteenth planet at the time of its discovery! As mentioned earlier, the Titius–Bode formula predicted a planet between Mars and Jupiter, at 2.8 AU from the Sun. Hectic attempts were made to locate it. In 1801, 20 years after the discovery of Uranus and 45 years before the discovery of Neptune, Giuseppe Piazzi discovered a planet located at about 2.7 AU from the Sun, almost exactly at a distance that the Titius–Bode formula predicted. But the size of the object was estimated to be very small as compared to even Mercury. It was named as Ceres after the Roman goddess of harvests. William Herschel, the discoverer of Uranus, and a few other astronomers were reluctant to consider Ceres as a planet because of its small size. Herschel called it an asteroid, a new class of objects in the solar system. But most of the astronomers at that time considered it to be a planet. So Ceres was the eighth planet in the solar system. Next year, in 1802, Heinrich Olbers discovered another object, even smaller than Ceres, located at a distance of about 2.6 AU from the Sun, very near to the orbit of Ceres. It was named as Pallas and was considered to be the ninth planet in the solar system. Just after 2 years of the discovery of Pallas, Karl Harding discovered another planet, Juno, in the vicinity of Ceres and Pallas. This was the tenth planet in the solar system. But the number ten lasted for just a few years. In 1807, Olbers discovered another planet Vesta at the same vicinity. So the number of planets increased from 7 to 11 within a span of 6 years. For the next four decades, 11 planets were known, and according to their distances from the Sun, they were (1) Mercury, (2) Venus, (3) Earth, (4) Mars, (5) Vesta, (6) Juno, (7) Pallas, (8) Ceres, (9) Jupiter, (10) Saturn, and (11) Uranus.

Thirteen Planets for a Year and Eight Planets for Eight Decades

At the end of 1845, just few months before the discovery of Neptune, Karl Ludwig discovered another planet located at a distance of 2.7 AU from the Sun, in the vicinity of Juno, Vesta, Ceres, and Pallas. This was named as Astraca and it was the fifth member of such small objects orbiting very near to each other. So with the discovery of Neptune in 1846, the total number of planets in the solar system increased to 13. But the size of Neptune as compared to the smaller planets Ceres, Pallas, etc. again initiated debate regarding the status of these small objects. So, by 1847, these five smaller objects Ceres, Pallas, Juno, Vesta, and Astraca lost their status as planets and were termed as minor planets or asteroids or planetoids. As a consequence the number of planets went back to seven plus Neptune, that is, a total of eight.

During the nineteenth century, a few astronomers suggested the existence of the ninth planet—a planet between Mercury and the Sun. It was named as Vulcan. The suggestion was put forward to explain the excess orbital precession of Mercury (known as perihelion precession). Subsequently, a few professional as well as amateur astronomers claimed to have observed transits of the Sun by this hypothetical planet. But it was never confirmed. Most probably a sunspot was mistaken as the shadow of the planet on the Sun. The excess orbital precession of Mercury was later on explained by Albert Einstein's famous "General Theory of Relativity."

It is interesting to note that from 1847 to 1930, the number of solar planets were eight, just the same to what it has become after 2006. Pluto was discovered by Clyde Tombaugh on 18 February 1930, and it enjoyed the status of the ninth planet in the solar system for the next 76 years. So, there were eight planets for about eight decades and there were nine planets for another eight decades.

Finally Pluto Is Not a Planet: End of Controversy?

So, the number of planets changes depending on the change in our knowledge about them. The planet Pluto was discovered in 1930. However, its mass was not known for more than four decades. It was thought that Pluto was as massive as the Earth. Pluto's main satellite Charon was discovered only in 1978. During the late 1970s, it has been realized that Pluto is about six times lighter and 1.5 times smaller than the Moon. On the other hand, Jupiter, the largest planet in the solar system, is about 146,000 times heavier than Pluto. The debate whether or not Pluto should belong to the group of celestial objects known as planets arises during the late 1990s after the discovery of a large number of planets outside the solar system. Some of these planets orbiting other stars are ten times heavier than Jupiter. Subsequently, the debate intensified when many small objects were discovered beyond the orbit of Neptune and at the extreme edge of the solar system. This region is known as the

Kuiper belt. All these objects have masses comparable to or even more than the mass of Pluto. Importantly, Pluto is also a Kuiper belt object. In 2003, two such objects—2003 UB313 (initially named as Xena) and Sedna—were discovered by Michael Brown and his group. It is estimated that 2003 UB313 is at least 25 % larger than Pluto and Sedna is about half the size of 2003 UB313. The situation became similar to the discoveries of Ceres, Pallas and other asteroids during the early seventeenth century. So, if Pluto is considered to be a planet, then all of these newly discovered objects should also be planets and the number of solar planets could continue to increase with the new discoveries in future.

Under this situation, the International Astronomical Union was compelled to form a committee consisting of eminent scientists from several fields to determine an appropriate definition of planets. Finally, on 24 August 2006, the 26th General Assembly of the International Astronomical Union passed resolution 5A by voting which states the following:

The IAU therefore resolves that planets and other bodies in the *solar system* except the *satellites*, be defined into three distinct categories in the following way:

1. A "*planet*" is a celestial body that (a) is in orbit around the Sun, (b) has sufficient mass for its self-gravity to overcome rigid body forces so that it assumes a *hydrostatic equilibrium* (nearly round) shape, and (c) has *cleared the neighbor hood* around its orbit.
2. A "dwarf planet" is a celestial body that (a) is in orbit around the Sun, (b) has sufficient mass for its self-gravity to overcome rigid body forces so that it assumes a hydrostatic equilibrium (nearly round) shape, (c) has not cleared the neighborhood around its orbit, and (d) is not a satellite.
3. All other objects except satellites orbiting the Sun shall be referred to collectively as "*small solar system bodies.*"

In the footnote of the resolution, Pluto is mentioned as a dwarf planet which is different from other planets.

Quite obviously, the above resolution generated controversies. Nevertheless, IAU has been very careful in the scope of the definition and emphasizes that the above definition applies to the solar system planets only. The extension of the definition of the solar planets to the planets outside the solar system would have generated further controversy and confusion. This will be clear in the subsequent chapters of this book.

So, right now we have a total of eight solar planets. According to their distances from the Sun, they are Mercury, Venus, Earth, Mars, Jupiter, Uranus, and Neptune. There are five recognized dwarf planets—Ceres, Pluto, Eris, Makemake, and Haumea. The dwarf planets that orbit beyond the orbit of Neptune are called trans-Neptunian objects or plutoids. Except Ceres, all other dwarf planets are plutoids. Ceres, first considered as a planet and then an asteroid, is now considered as a dwarf planet and it is located in the asteroid belt, between Mars and Jupiter. The number of dwarf planets will certainly increase with new discoveries. But the confusion, controversy, doubt, and debate will continue.

Chapter 3
Our Neighborhood: The Solar Family

> *The unquiet republic of maze*
> *Of Planets, struggling fierce towards*
> *heaven's free wilderness.*
>
> –P. B. Shelley
> *(In Prometheus Unbound, 1820)*

The atmospheres or climates and the dynamics of all the planets in the solar system are governed solely by their parent, the Sun. We exist because of the Sun. We survive because of the Sun. Our world is blessed by the Sun. Therefore, before we discuss the climates and the physical properties of the solar planets including our own world, the Earth, we should have a fairly good idea about the Sun, the nearest star that supplies energy to each and every members of the solar system. On the other hand, without a fair knowledge on the Sun and the solar planets, we may not be able to appreciate the diversity of planets around other stars. Since the chemical composition of a planet is almost the same to that of the parent star, it should also be easier to characterize a planet or a planetary system around a star similar to the Sun.

Sun: The Parent of the Solar Family

In the whole solar system, the Sun is the only source of continuous energy for any physical processes including life on the Earth. It plays a central role in the formation of the planets around it, in governing the motions of the planets, in determining the climates of the planets, and in protecting the planets as well as life on the Earth. We have already mentioned that the Sun is classified as a yellow dwarf and it is a G2V-type star in the spectral classification schemes. However, among the 200 billion stars in the vast Milky Way Galaxy, the Sun is an ordinary, mediocre, and middle-aged star.

The Sun is a second-generation star in the sense that except hydrogen and helium, all other materials inside it were synthesized inside a first-generation star and during its explosion as a supernova. When a first-generation star exploded, the heavy elements—carbon, oxygen, neon, iron etc.—mixed up in the interstellar cloud and the gravitational collapse of such cloud formed our Sun. About 73.5 % of its total mass comprises of hydrogen and about 24.8 % of the total mass is contributed by helium. The remaining 1.7 % of the mass which is equivalent to 5,600 times the mass of the Earth consists of other elements that were synthesized inside a first-generation star.

It takes 8 min 20 s for the sunlight to reach us, while light from the star closest to the Sun, Alpha Centauri, takes a long four years to reach us. Actually, Alpha Centauri belongs to a triple star system. Alpha Centauri A and Alpha Centauri B are orbiting each other, while a third star, Proxima Centauri, is orbiting the Alpha Centauri binary system called as Alpha Centauri AB. Most of the stars in our galaxy are either in a binary or in a multiple star system. But our Sun is an isolated star considered as a field star. The Sun is about 109 times larger than the Earth with a diameter of about 1.4 million km. It is so voluminous that about 13 million Earths can easily be placed inside the Sun. However, the mean density of the Sun is about 1.4 g/cm^3, whereas the density of the Earth is about 5.5 g/cm^3. The Sun has a differential rotation around its own axis. It takes about 26 Earth days at the equator and 36 Earth days at the poles to rotate around its own axis. The Sun consists of about 99.86 % of the total mass of the solar system, and it is about a hundred times heavier than Jupiter. It has a very active magnetic field which causes sunspots and solar activities. The polarity of the magnetic field changes in a cycle of 11 years, and this process is called the solar cycle. The number of sunspots, flares, winds etc. varies during this cycle from a minimum to a maximum. The sudden outburst of solar flares causes disturbances in our telecommunication system, and the solar wind affects the magnetic field of the Earth.

The entire star is made of extremely hot gas which is often called as plasma—the fourth state of matter. Its structure can be divided into six regions. The innermost region known as the core has a temperature of about 15 million degree Celsius and is under tremendous pressure. Such condition enables nuclear fusion of hydrogen atoms. This nuclear process supplies the energy continuously. The region above the core is called the radiative zone where the energy is transferred through radiation process. The region above the radiative zone is convective. In this region, the up and down movements of the hot plasma transfer the energy just like boiling water. The photosphere and chromospheres of the Sun are situated above the convective region. The photospheric temperature of the Sun is about 5,504 °C. The temperature thereafter increases again to a million degree in a much diluted region called the corona. The heating of the corona is caused by the release of magnetic field energy. The sunspots are the cooler regions in the photosphere with a temperature of about 4,000 °C. The rate of energy production inside the Sun or the power of the Sun is increasing slowly. The Sun was born about 4.6 billion years ago. At that time its brightness was about 30 % less than the present brightness. Every one billion years, the brightness of the Sun increases by 10 %. Therefore, after a billion years from now, the Sun will be so hot that all the liquid water on the surface of the Earth will start boiling and become vapor. Besides the visible light, the Sun also emits X-rays and ultraviolet rays.

The Sun is composed of various elements in different proportion, and the astronomers call it the solar abundance. The chemical composition of the Sun is estimated from the analysis of the spectrum of the Sun. This spectrum is originated at the photosphere and at the chromosphere of the Sun. The chromosphere is located just above the photosphere. The photosphere is defined as the surface below which the Sun is opaque to visible light. Therefore, in principle, we can estimate the chemical composition of the visible atmosphere of the Sun from the spectrum. But it is considered that the chemical composition of the atmosphere represents the entire star. About 67 elements have been detected from the spectrum of the Sun. Besides hydrogen and helium, it contains about 1 % oxygen and about 0.4 % carbon. In fact the carbon to oxygen ratio in the solar system is about half and that is why in our planet and around our neighborhood, we see a large number of oxide compounds—compounds made of oxygen atoms such as water, carbon dioxide, sulfur dioxide, silicon oxide, etc. The amount of other elements in the Sun such as nitrogen, silicon, and magnesium is about one tenth of the amount of oxygen present in it. Neon, iron, sulfur, etc. are present in a very small amount, about one tenth of the amount of carbon present.

Climate of the Solar Planets

The stars are born out of the fragmentation and gravitational collapse of huge molecular clouds, while planets are born out of the proto-planetary disk around a newly born star. However, as we will see at a later stage, the definition of planets on the basis of their formation or birth history has in recent years emerged out to be ambiguous. Nevertheless, it is now well established that at least the solar planets are born out of the proto-planetary disk through accretion and accumulation of gas, dust, and small celestial bodies. This is one of the reasons why IAU has been careful in resolving the definition of only solar planets.

Since both the parent star and the planets around it are formed out of the same material, the material compositions of the parent star and the planets are almost the same. During the proto-planetary stage, the lighter elements fall onto the star and so usually the planets; in particular the inner planets are slightly richer in heavy elements. The stars have their own source of light, whereas the planets reflect the starlight. The light that we see from a star or from a planet emerges from a thin region situated at the uppermost part of the object and is known as the atmosphere. Below the atmosphere, the object is opaque to light. Light travels through the atmosphere and undergoes various physical processes such as absorption, scattering, re-emission, etc. before it reaches to us. Therefore, the light we receive from a star or from a planet carries the information of the stellar or the planetary atmosphere. Astronomers decode this information to determine the chemical composition and physical properties of the star or the planets. The heavier the star, the brighter it is because it burns more fuel at a given interval of time. There is a relationship between the brightness or luminosity and mass of a star. The brightness or luminosity of a star also tells the age because it changes with the age of the star. Since the heavier and so brighter stars burn their fuel more rapidly, they live shorter. The age of a planet is almost the same to its parent star. Thus, the age of a planet or a planetary system is determined from the age of the parent star. The surface gravity and the temperature of a star can be determined from the spectrum. For a binary star system, the mass of each star can also be determined from the orbital period and the distance between the two stars by using Kepler's third law. Surface gravity determines the pressure in the atmosphere at different heights.

Since a planet does not have its own source of energy (unless it is quite young, less than a hundred million years old), its climate is determined by the amount of energy its atmosphere receives from the parent star. The amount of energy that reaches the planet per second from the star depends on (1) the power of the star, i.e., the amount of energy the star emits per second, and (2) the distance between the planet and the star. A small fraction of the energy or radiation cannot penetrate into the atmosphere and so scatters back into the space. The remaining energy enters the atmosphere and gets absorbed, scattered, and re-emitted in different wavelengths. The ratio between the reflected light and the incident starlight at the surface of the planet is called the albedo which is an important measurable quantity. The albedo depends on the thickness, depth, and chemical properties of the material present in the atmosphere. The albedo varies from zero to one. For example, if the albedo is one, then all the light incident on the surface of the planet is reflected back to the space. If it is zero, then all the incident starlight is absorbed by the planet. Thus, the surface temperature of a planet is determined by the luminosity or the power of the star, the distance between the star and the planet, and the reflectivity of the planetary surface. On the other hand, the composition of the atmosphere is determined by the amount of energy that is absorbed by the planetary atmosphere. For example, if the brightness of the Sun is increased or if the Earth moves closer to the Sun, the surface temperature of the Earth would become so high that water would evaporate from it. Similarly, the methane ice would have melted and become gaseous had Jupiter moved to the position of the Earth.

The solar system has a characteristic in its distribution of planets which is unique among all the planetary systems discovered till date. The smaller and rocky planets are closer to the Sun and also known as the inner planets, while the giant gaseous planets are far away from the Sun. The difference in the sizes of the solar planets can be visualized from Fig. 3.1. So far, not a single extra-solar planetary system is detected which shows the same characteristic. As we shall see later on, the presence of giant planets such as Jupiter in fact helps in protecting life in the inner planet Earth. Now we shall briefly discuss some of the most important physical as well as atmospheric properties of the solar planets. Let us first begin with our own world, the Earth.

Fig. 3.1 A comparison (not in actual scale) of the sizes of the solar system planets (Credit: Efrain Morales Rivera, Jaicoa Observatory, Aguadilla, Puerto Rico; with permission)

The Earth

The Earth is the third member of the solar system according to its proximity to the Sun. It is about 150 million km away from the Sun. The distance between the Earth and the Sun is called 1 Astronomical Unit or 1 AU which is equal to 149,598,262 km. It is convenient to refer the distance of all the planets including the solar planets from their parent stars in term of AU. The Earth is a rocky planet with an atmosphere consisting mainly of nitrogen and oxygen gas, comparatively less amount of carbon dioxide, argon, etc., and it has plenty of liquid water. By volume, atmosphere of the Earth has 78 % nitrogen, 21 % oxygen, about 1 % argon and water vapors, and a very little, 0.04 % carbon dioxide. The upper atmosphere of the Earth contains a thin layer of ozone. Ozone is either not present at all or very less in the atmospheres of other planets of the solar family. We shall discuss about the formation of the ozone layer in a later chapter.

All the inner planets including the Earth have very high density because of the presence of a rocky core. Although the diameter of the Earth is as small as about 12,742 km at the equator, its density is about 5.5 g/cm^3. The Earth is the densest planet in the solar system. It is also the largest planet among the inner rocky planets of the solar family. The Earth has an active geology with volcanoes and earthquakes that also influence the Earth's atmosphere by reprocessing various chemical compounds in regular intervals. The Earth rotates around the Sun at a speed of about 107 km/h in almost a circular orbit. The surface temperature of the Earth varies from about $-88\,°C$ to $58\,°C$ and the mean temperature is $15\,°C$. Therefore, water in all the three states— solid, liquid and gas in the form of ice, liquid water, and water vapor or water cloud—is present over the surface of it. Carbon dioxide and water in the Earth's atmosphere reflect a fraction of sunlight back into the space and also keep the Earth warm by trapping the re-emitted solar energy. In the absence of carbon dioxide and water, the Earth could have been much cooler. This is known as Greenhouse effect. In other planets methane plays the same role.

The Earth has a magnetic field which is dipolar in nature. It's like a bar magnet. The North Pole of the magnetic field is located at the geographic South Pole, and the South Pole of the magnetic field is located at the geographic North Pole. However, the line joining the magnetic poles or the magnetic axis is oriented at an angle of $11°$ with respect to line joining the geographic poles. In other words, the magnetic axis is presently tilted by $11°$ with respect to the spin axis of the Earth. The magnetic polarity changes with time, and a complete reversal takes about 50 million years. The magnetic field is also gradually losing its strength. The magnetic field does not allow the energetic charged particles of the solar wind to enter into the Earth's atmosphere. The charged particles can enter only through the magnetic poles causing spectacular aurora.

The rotation axis of the Earth is tilted by an angle of $23.4°$ with respect to the orbital plane, and this tilt or obliquity causes the seasonal variation. Earth is not perfectly spherical in shape—the equatorial radius is slightly larger than the polar radius. Finally the Earth has a large satellite, the Moon which is tidally locked with the Earth, and so the rotation of the Moon around the Earth is synchronized with the spin around its own axis. Most probably the Moon was created about 4.5 billion years ago when a Mars-sized object collided with the Earth. We shall discuss about the interior of the Earth when we compare it with the Super-Earths that are discovered recently. However, there is no Super-Earth in the solar system.

Mercury

Since Pluto is no more considered to be a planet, Mercury has got the record of being the lightest planet in the solar system. It is about 20 times lighter than the Earth. However, it is the second densest planet in the solar system with a density of $5.4\,g/cm^3$, just a little less than the density of the Earth. Mercury is the closest planet to the Sun, just 58 million km away from it, and hence, it is extremely hot. The surface temperature of Mercury varies from $-173\,°C$ to $430\,°C$. Mercury is also the smallest planet in the solar system with a diameter of about 4,900 km. Mercury's one year is just about 88 days of the Earth, but one day of Mercury lasts as long as 58 and a half days of the Earth. This means Mercury rotates exactly thrice around its own axis during two orbits around the Sun. This is due to the tidal interaction of the Sun and is known as spin–orbit coupling. Mercury should have been rotating faster in the past but was slowed down substantially by the Sun. The orbital precession of Mercury was used to prove Einstein's General Theory of Relativity.

Much about Mercury was first known by the space mission Mariner 10 which orbited the planet as close as a few hundreds of kilometers. Mercury has no atmosphere because of its close proximity with the Sun and because of very low surface gravity. The surface gravity of Mercury is about 38 % of the Earth's surface gravity. So, either the atmosphere had evaporated completely or was not even formed. The surface of Mercury is similar to that of our Moon, but the interior of the planet is very similar to the Earth. Mercury has a large number of craters and basins. In the absence of volcanic activities or plate tectonics or atmospheric effect, the craters and basins that were created at the early age of the planet by the bombardments of asteroids remain the same.

Unlike the Earth's rotation axis, Mercury's rotation axis is not tilted and it is perpendicular to the orbital plane. Still it has Earth-like seasonal changes in the surface temperature at different longitudes. This is due to the elliptical orbit and the spin–orbit coupling. Due to spin–orbit coupling, Mercury's noon lasts longer when the planet is at its perihelion (the closest approach to the Sun). In fact an observer on Mercury when it is at the perihelion would see that the Sun stops moving for some time and then moves retrograde for a few days. Mercury has an Earth-like dipole magnetic field. But the magnetic field strength is about 1 % of the Earth's magnetic field strength, and the magnetic axis is aligned with the rotation axis. Since Mercury is very dense, it is considered that the planet has a liquid iron core similar to the core of the Earth. However, Mercury rotates slowly and hence the magnetic field might not have been originated by the geological effect. It is possible that Mercury's magnetic field is induced by the deposition of charged particles carried by the solar wind. Mercury has no moon of its own.

Venus

Venus is the second planet of the solar system at a distance of about 108 million km from the Sun and moving in an almost circular orbit around it. Venus completes one orbit around the Sun in 225 days which means a year of Venus is 225 days of the Earth. Venus is considered to be the twin sister of the Earth as both planets have similar sizes, masses, densities, compositions, and surface gravities. However, Venus rotates quite slowly around its own axis. It takes 244 Earth days for Venus to rotate around its own axis, making it the slowest rotator among all the solar system planets. But one day of Venus is not equal to 244 Earth days! While all the solar system planets rotate around their own axis in counterclockwise, Venus rotates in the opposite direction around its own axis—from the East to the West. Therefore, in Venus the Sun appears to rise in the West. Because of the retrograde rotation, it takes about 117 Earth days from one sunrise to another.

Venus has a very thick and heavy atmosphere. It is about 90 times heavier than the atmosphere of the Earth. Therefore, the atmospheric pressure at the solid surface of the planet is extremely high. On the other hand, because of this dense atmosphere, a large amount, about 90 % of the solar radiation, gets reflected back from the upper atmosphere. As a consequence, Venus should be cooler. But contrarily, it is the hottest planet in the solar system although it is not the closest one to the Sun. This is due to the fact that the dense atmosphere absorbs the solar heat and a runaway Greenhouse effect, much more effective than that in the Earth, makes Venus very hot. The mean surface temperature of Venus is about 462 °C. By volume, Venus has about 96 % carbon dioxide, 3 % nitrogen, and a very little amount of argon, water vapor, helium, neon, sulfur dioxide, etc. Venus could have water oceans in the past, but water was dissociated and subsequently evaporated by the intense heat and photoionization by ultraviolet rays of the Sun. Similar to the Earth's atmosphere, thunderstorms and lightning in Venus's atmosphere have been detected. The top layer of Venus's cloud travels the whole planet in just four Earth days due to wind moving with a speed of about 360 km/h—about 60 times faster than the rotation of the planet itself. At the bottom of the atmosphere, however, the wind speed is quite slow, just a few kilometers per hour. The lightning detected in Venus is due to clouds of sulfuric acid and not due to water clouds. Recently, trace of ozone is detected in the upper atmosphere of Venus. However, unlike the Earth wherein ozone is formed mainly due to bio-chemical mechanisms, in Venus, ozone is formed purely by chemical processes—through the decomposition of carbon dioxide.

Venus has no geomagnetic field because it rotates very slowly. Also Venus has a metallic iron core without much circulation that is essential to generate the geomagnetic field. Therefore, charged particles carried by the solar wind affect the atmosphere. Its rotation axis is perpendicular to the orbital plane and so there is no longitudinal variation on the surface temperature. The surface of Venus is full of large craters. Venus has thousands of volcanoes with diameter ranging from 1 km to 250 km. There are at least 170 volcanoes which are 100 km wide. The volcanic activity in Venus is very high and the lava from these volcanoes smoothen the surface of Venus. As a consequence, the surface features of Venus are mainly determined by the volcanic activities. Venus too has no moon.

Mars

Mars is the fourth and the last inner planet of the solar system. The mean distance of Mars from the Sun is about 228 million km or one and a half AU. It takes almost two years (687 Earth days) for Mars to orbit the Sun in an elliptical orbit. Mars is about half the size of the Earth and ten times lighter than the Earth. However, it has only slightly longer day. One Martian day is 24 h and 37 min. The surface gravity of Mars is almost the same to that of Mercury. Anybody weighing 100 kg on the Earth would have a weight of just 38 kg on Mars. Mars has an atmosphere which is much thinner than the Earth's atmosphere. The atmospheric pressure at the surface of Mars is 200 times less than that on the Earth. Martian atmosphere has some similarity to the atmosphere of the Earth, but unlike the Earth's atmosphere, Mars has about 95 % carbon dioxide, about 2–3 % nitrogen, 1.6 % argon, 0.13 % oxygen, and a little amount of water vapor in the upper atmosphere. The polar regions of Mars are covered by water ice. The amount of ozone in Martian atmosphere is a thousand times less than that in the atmosphere of the Earth. This ozone is produced through the decomposition of carbon dioxide. Therefore, in the absence of an ozone layer, the strong ultraviolet rays from the Sun dissociate the carbon dioxide into the poisonous carbon monoxide gas. At the same time, water vapor is dissociated into hydroxyl and hydrogen molecules.

The rotation axis of Mars is tilted by 25° with respect to the orbital plane. Therefore, similar to the Earth, Mars too have seasonal variation in its climate. But the highly elliptical orbit makes the climate extreme. When the planet is farthest to the Sun (at the aphelion), it receives about 40 % less radiation than the amount it receives when it is closest to the Sun. The reflectivity or the albedo of Mars is 0.2 which means 20 % of sunlight received is reflected back by the atmosphere. This is almost the same to that of the Earth. Although Mars has a huge amount of carbon dioxide in its atmosphere which should cause higher reflectivity, the thin atmosphere makes the albedo of Mars slightly lower than that of the Earth. However, Mars cannot trap much of the incoming heat because of the low density and absorption capacity of the atmosphere. There is practically no Greenhouse effect in Mars. The surface temperature of Mars varies from $-140\ °C$ at the polar regions during the winter to $35\ °C$ during the summer at the equator. The elliptical orbit of Mars makes its climate interesting. The northern hemisphere of Mars has a shorter winter and is comparatively warmer than the southern hemisphere. But the summer lasts longer and is cooler as compared to that in the southern hemisphere. When Mars is nearest to the Sun, its southern hemisphere is tilted towards the Sun. This makes a short but hot summer in the southern hemisphere and a short winter in the northern hemisphere. On the other hand, when the planet is farthest to the Sun, the northern hemisphere is tilted towards the Sun, making both the summer and the winter longer but cooler. Each session on Mars is double in duration as compared to that on the Earth because one Martian year is almost double to one Earth year.

Because of low surface gravity and strong wind, almost all the time there is strong dust storm, both globally and locally. The dust cloud extends even up to 40 km above the surface, preventing sunlight to enter. The dust storm in Mars is the largest in the entire solar system and can cover the entire planet for a long time. The dust is composed mainly of silicate oxide or sand. Mars has large craters and volcanic mountains. Although Mars is less massive and hence it attracts less number of cosmic heavy materials such as asteroids and meteoroids, it is closer to the asteroid belt and so it has suffered heavy bombardment of asteroids. However, small meteoroids are burnt in the atmosphere before they hit the surface. Mars has about 45,000 craters with diameters ranging from 1 km to a few hundred kilometers. The craters which are greater than 1 km but less than 60 km in radius are named after various places on the Earth. Craters larger than 60 km in radius are named after great scientists, geologists, artists, writers etc. The largest crater, the

Hellas Planitia, is about 2,300 km wide and located in the southern hemisphere which has more craters than the northern hemisphere. At the same time Mars has several mountains and volcanoes. The mountain called Olympus Mons is about 27 km high, more than three times the height of the tallest mountain on the Earth, Mt. Everest. Olympus Mons is actually a volcano with a diameter of about 600 km. However, all the volcanoes were active only during the early age of the planet. Mars has valleys, channels, canyons, etc. as well. The crust of the Mars is much thicker than that of the Earth. The average thickness of the crust is about 40 km which prevents plate tectonics. As a consequence, Mars is geologically dead and hence cannot reprocess the chemical compounds that dissolve into the Martian rocks. Although the core of Mars consists of iron sulfide, it must not be in liquid form because in that case, the sufficiently fast rotation of the planet would have given rise to geomagnetic field. But Mars has no magnetic field. As a consequence, charged particles carried by the solar wind penetrate the whole atmosphere and ionize the atmosphere. Both the polar caps of Mars are covered by water ice and frozen carbon dioxide. It is believed that about 3.5 billion years ago, Mars had plenty of water in its surface. The water-driven erosion of Martian surface and the detection of minerals such as hematite and gypsum are the evidence for it. However, the present low temperature, low atmospheric pressure, and strong ultraviolet rays do not allow water to exist in liquid form. Further, the tilt in the rotation axis of Mars which causes the seasonal variation in the climate alters drastically and chaotically during a million years. This gives rise to dramatic change in the climate. This change in the tilt of the rotation axis is produced by the gravitational effect of Jupiter and Saturn.

Fig. 3.2 Mars and its two moons, Deimos and Phobos (Credit: Efrain Morales Rivera, Jaicoa Observatory, Aguadilla, Puerto Rico; with permission)

Mars has two satellites, Deimos and Phobos, which are much smaller than the Moon, a few kilometers in size (Fig. 3.2). These non-spherical satellites may be asteroids captured by Mars.

Jupiter

The largest and the heaviest planet in the solar system is Jupiter. Together with the other three gas giant planets—Saturn, Uranus, and Neptune—Jupiter is grouped into the outer planets separated from the inner planets by the asteroid belt which contains thousands of smaller rocky bodies. The outer planets are also called Jovian planets. Jupiter is about 778 million km or about 5.2 AU away from the Sun. How large and heavy is Jupiter? It is about 320 times heavier than the Earth and a thousand times lighter than the Sun. Jupiter is about 2.5 times heavier than all the other seven planets in the solar system put together. It is about 11 times larger than the Earth in diameter. This planet is so large that about 1,300 Earths can be fitted inside it. However, its density is one fourth of the density of the Earth. It is also rotating very fast around its own axis. One Jupiter day takes just about ten hours. Jupiter is the fastest rotator among all the planets in the solar family. Because of this, Jupiter departs significantly from a spherical shape. It has a bulge at the equator. The equatorial radius is about 7 % longer than the polar radius. The rotation axis is almost perpendicular to the orbital plane. Jupiter rotates around the Sun in a slightly elliptical orbit and it takes slightly less than 12 Earth years to make one orbit around the Sun. Jupiter is so heavy that it causes significant wobble in the Sun. It has no solid surface. The entire planet is made of gas. Its core is probably liquid because of high pressure. Jupiter's reflectivity or albedo is very similar to that of the Earth. The mean surface temperature of Jupiter is about -148 °C which is slightly higher than its equilibrium temperature (-110 °C) that can be derived from the incident solar heat. Therefore, Jupiter has an internal source of energy. This is probably due to the fact that Jupiter is still shrinking slowly and thus emitting gravitational potential energy. The central region of Jupiter may have temperature ranging from 15,000 to 35,000 °C, and the pressure at this region may be a hundred million times the Earth's atmosphere.

The atmosphere of Jupiter is mainly composed of molecular hydrogen (about 90 %) and helium (about 9 %). Apart from hydrogen and helium, the atmosphere has little amount of water, methane, and ammonia. Jupiter has patchy clouds with cyclonic weather. Because of very low temperature, the cloud mainly consists of ammonia vapor. Water, methane, and ammonia ices are present in the polar region. Jupiter has several spots of different sizes, colors, and lifetimes. The color of the spots varies according to the color of the chemical composition of the clouds. The famous red spot is 24,000 km long and 11,000 km wide which means the size of the red spot is about three times the Earth's diameter. The red spot is actually a gigantic storm very similar to a hurricane on the Earth. It is more than 400 years old.

Although Jupiter is a gaseous planet, due to huge pressure, most of the interior of Jupiter is in liquid form. It may have a small solid core at the center. Jupiter has a very strong magnetic field. The magnetic pole strength of Jupiter is about 20,000 times greater than that of the Earth. However, since the field strength reduces with the cube of the distance from the magnetic axis and Jupiter is about 12 times larger in radius than the Earth, the magnetic field at the surface is about 14 times stronger than that of the Earth. The magnetic field is extended up to about 3 million km from the planet. This is resulted from the electrical current in the core that mainly consists of rapidly spinning "metallic hydrogen." Hydrogen under normal condition is nonmetallic, but under extreme pressure inside the planet, hydrogen behaves like metal. The solar wind that carries charged plasma strongly affects the shape and size of the magnetosphere of Jupiter. Under strong solar wind, the magnetosphere shrinks substantially.

Jupiter has a few thin rings around it. The main ring is about 30 km thick and 6,500 km wide. The rings are most probably made of small dust particles. Jupiter has at least 67 moons orbiting it. The largest moon Ganymede has a diameter of 5,262 km—almost the size of Mercury—and it is two times heavier than our Moon. The other giant moons or satellites of Jupiter are Callisto, Io, and Europa (Fig. 3.3).

Fig. 3.3 Jupiter and two of its moons—at the *top* of the planet is Callisto and at the *bottom right* is Ganymede, the largest satellite of the solar system (Credit: Efrain Morales Rivera, Jaicoa Observatory, Aguadilla, Puerto Rico; with permission)

Saturn

Saturn is the second largest as well as the second heaviest planet in the solar system but the least dense planet among the eight solar planets. Saturn is about 95 times heavier and about 9.5 times larger than the Earth. Its density is less than even water, and so Saturn would float in water. Saturn takes 29½ years to orbit the Sun in an almost circular orbit. It is about 1,433 million km or about 9.5 AU away from the Sun. Saturn takes about ten and a half hours to rotate around its own axis which means one day of Saturn is slightly less than half a day of the Earth. The rotation axis is tilted by 26.7° with respect to the orbital plane. This planet is very similar to Jupiter in structure—it has a liquid core covered by a layer of liquid hydrogen. Saturn too has geomagnetic field due to fast spin, but the surface magnetic field is much weaker than that of Jupiter. Because its density is very low but it rotates quite fast around its own axis, Saturn has the maximum oblateness or departure from a spherical shape. The equatorial radius is about 10 % longer than the polar radius.

Similar to the atmosphere of Jupiter, the main materials in Saturn's atmosphere are molecular hydrogen and helium. The atmosphere contains about 96 % molecular hydrogen and almost 4 % helium. Other gases such as methane (0.4 %) and ammonia (0.01 %) are also present in the atmosphere. The albedos of Saturn and Jupiter are the same, 0.34 which means both the planets reflect about 34 % sunlight

Fig. 3.4 Saturn with its spectacular rings and four satellites (Credit: Efrain Morales Rivera, Jaicoa Observatory, Aguadilla, Puerto Rico; with permission)

back to the space. The Earth too has the same reflectivity. The mean surface temperature of Saturn is about $-140\ °C$, slightly higher than it should be due to the solar heat received. Saturn too has a cloud of ammonia but the wind speed is much higher than that in Jupiter. It too has a spot—the Great White Spot which extends around the whole planet. This spot is due to a huge storm. However, this is a short-lived storm which occurs every year in Saturn. Saturn is well known for its spectacular rings which are composed mainly by dust. At least 62 satellites of various sizes orbit around Saturn. The largest one is Titan. Other remarkable satellites are Dione, Mimas, Enceladus, Tethys, Rhea, etc. (Fig. 3.4).

Uranus

Uranus is the first planet discovered by modern-day astronomy as discussed in the previous chapter. It was discovered by William Herschel on the 13 March 1781. Actually, Uranus was observed before by many people, but it was mistaken as a star. At present it is the seventh planet from the Sun. The average distance of Uranus from the Sun is about 2.9 billion km or 19 AU, and so it takes 84 Earth years for Uranus to make one orbit around the Sun. Uranus is smaller than Saturn. It is about four times larger but about 15 times heavier than the Earth. Therefore, this planet is the fourth heaviest and the third largest among all the solar planets. Uranus has a density almost similar to that of Jupiter, about 1.27 g/cm^3. The surface gravity of both Saturn and Uranus is slightly less than that of the Earth. It takes about 17½ h for Uranus to complete one rotation around its own axis. So, is the day in Uranus as long as 17½ h? Quite interestingly it is not. Uranus has a unique feature—its spin axis is tilted almost parallel to the orbital plane. While the rotation axis of the other solar planets are almost perpendicular to the orbital plane and so they rotate like slightly tilted spinning tops around the Sun, Uranus looks like a ball rolling on its side. The tilt angle or obliquity is 97.7°. Obviously this affects the climate of the planet severely by giving rise to extreme seasons, each of which has duration of about 21 years. One of the poles is directed towards the Sun, and therefore, the duration of a day or a night of one hemisphere is as long as 42 Earth's year. Uranus too has a geomagnetic field and the magnetic axis is tilted by 58.6° with respect to the rotation axis. The magnetic field is about 50 times stronger than that of the Earth. As a consequence, just like the other planets having magnetic field, the magnetosphere of Uranus too traps charged particles in belts. The magnetic field in Uranus is asymmetric—the surface magnetic field at the northern hemisphere is ten times stronger than that at the southern hemisphere. The average surface temperature of Uranus is about $-200\ °C$ or even less. Unlike Jupiter or Saturn, Uranus emits less heat than it receives and absorbs from the Sun. The atmosphere of Uranus is very similar to that of Jupiter. It has about 82 % molecular hydrogen, 15 % helium, and slightly more than 2 % methane. Uranus has patchy cloud of methane and so it looks bluish. The wind speed in Uranus is higher than the wind speed in Jupiter, but it is less than that in Saturn. Uranus has 13 rings around it and 27 moons

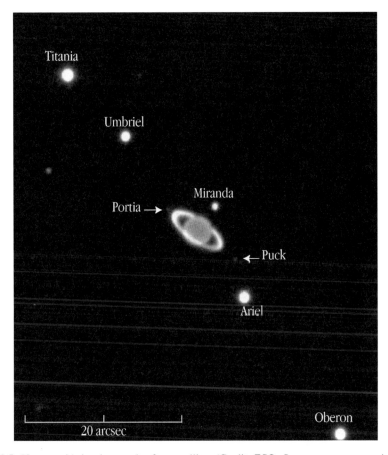

Fig. 3.5 Uranus with its rings and a few satellites (Credit: ESO. *Source*: www.eso.org/public/images/eso0237b)

of different sizes. The largest moons Titania and Oberon were discovered by William Herschel in 1787. Incidentally, Herschel himself discovered Uranus six years before. The other three large satellites of Uranus are Ariel, Umbriel, and Miranda (Fig. 3.5). Uranus has quite a few irregular satellites.

Neptune

According to the resolution passed by the General Assembly of the International Astronomical Union in 2006, Neptune is the last or the most distant planet in the solar system. Neptune was discovered by Johann Gottfried Galle on 23 September 1846. On average, it is about 4.5 billion km or 30 AU away from the Sun, and it takes about 165 years for Neptune to rotate around the Sun in an elliptical orbit. Neptune is just four times larger than the Earth. Its diameter is about 1,500 km shorter than the diameter of Uranus. But Neptune is heavier than Uranus. It is about 18 % more massive than Uranus and is about 17 times heavier than the Earth. The surface gravity of Neptune is about 17 % more than the surface gravity of the Earth. Neptune has a rocky core consisting of iron and magnesium silicate and a mantle of solid water, ammonia, and methane. Similar to Jupiter and Saturn, Neptune too has its own source of heat. The heat radiated by it is about 2.7 times more than the heat it receives from the Sun. However, Uranus does not have such a heat source, making it the coolest planet in the solar system. So in the solar system, the planet closest to the Sun is not the hottest one and the planet farthest to the Sun is not the coolest one. One Neptune day is about 16 h long. The rotation axis is tilted by 28.32° with respect to the orbital plane. Neptune too has a magnetic field very similar to that of Uranus but weaker in strength. Neptune's magnetic field is about 27 times stronger than that of the Earth. The geomagnetic axis is tilted by about 50° with respect to the rotation axis.

Neptune's albedo or reflectivity is slightly less than that of the Earth. The average surface temperature of Neptune is about -200 °C, similar to that of Uranus. But Neptune has nitrogen molecules in its atmosphere. In addition to nitrogen, it has about 83–85 % molecular hydrogen, 15–17 % helium, and about 1.5 % methane. The nitrogen molecules are most probably originated from its moon, Triton. Neptune has atmospheric cloud consisting of methane and hydrogen cyanide. It has several spots due to the presence of scattered clouds in its atmosphere. The Great Dark Spot which was a storm spinning counterclockwise had different shapes, locations, and orientations at different times. It was 12,000–7,400 km in length and 18,000–5,200 km in width. The storm had a speed of about 1,200 km/h. This spot or storm was detected by Voyager II. However, later on Hubble Space Telescope could not find it which implies that the same storm does not exist anymore. Neptune's wind can reach a speed as high as 2,500 km/h, making it the fastest in the solar system. Neptune has 5 rings around it and 14 moons, Triton being the largest among them which too has an atmosphere (Fig. 3.6). Triton was discovered by William Lassell in 1846 just a few weeks after the discovery of Neptune. The second moon Nereid was discovered in 1949 by Gerard P. Kuiper. Many of the other moons of Neptune were discovered by Voyager II spacecraft. Some of the moons of Neptune were discovered very recently by large ground-based telescopes.

Why Pluto Is Not a Planet?

Now that we have some information on the planets in the solar system, two questions arise immediately. The first one is why Pluto should not be considered as a planet. The second one is if there exists life beyond the Earth but within our neighborhood, the solar system. A group of objects having many common

Why Pluto Is Not a Planet?

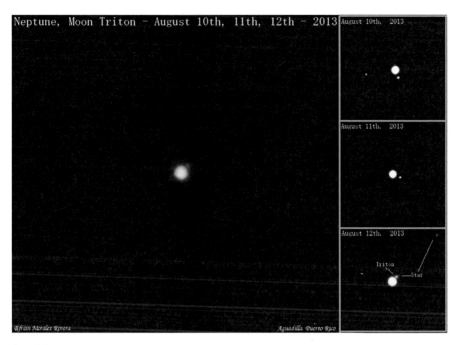

Fig. 3.6 Neptune and its satellite Triton (Credit: Efrain Morales Rivera, Jaicoa Observatory, Aguadilla, Puerto Rico; with permission)

properties are usually classified by the same name in order to make any member of them distinct from another group having different properties. For example, if we say Sirius is a star, we do not need to describe that it is a hot gaseous object which generates its own energy through nuclear burning at its central region etc. Similarly, if we say the Sun or the Pole star is a star, it immediately implies that both of them are similar to Sirius in nature. But we cannot say that Mars is a star as it does not resemble either the Sun or Sirius or the Pole star. Similarly, if we say that the Moon is a satellite, it would mean that this object rotates around a planet and not around a star. Therefore, the definitions are made for convenience. The main purpose of this classification is to make it easy to understand the basic properties of different celestial objects. It would be misleading and needs to add several exceptions if we say that the Moon is a planet. Now Pluto has some similarities with the other solar planets. It rotates around the Sun. It does not generate its own energy. Then what was the problem? The major problem was that nothing much was known about Pluto before 1978. Even the actual mass of Pluto was not known because the dynamical mass that was estimated from its effect on Uranus and Neptune was the sum of the mass of Pluto and all of its satellites. Charon, Pluto's satellite, was discovered only in 1978, and immediately it is understood that Pluto is lighter than even the Moon. If Charon rotates around Pluto, then it can be said that Pluto too rotates around Charon. In that case, Pluto could have been a binary planet, a unique feature in the solar system. However, Pluto is sufficiently massive so that it has a spherical shape due to its own gravitational force. The debate on Pluto's status as a planet intensified when objects of similar mass, some even more massive (such as Eris), were discovered in the region starting after Neptune's orbit to farther away known as the Kuiper Belt. Kuiper Belt is known to be a disklike region where objects much smaller than the Moon, e.g., the asteroids, are found and all of them rotate around the Sun. It extends from about 30 AU to 55 AU from the Sun. Kuiper Belt was predicted by Gerard P. Kuiper in 1950, 20 years after the discovery of Pluto. If we consider that all objects that rotate around the Sun are planets, then all the Kuiper Belt objects as well as the objects in the asteroid belt, millions in number, have to be considered as planets. But there is a minimum mass which makes the object spherical by its own gravitational force. So, if all those objects that rotate around the Sun and have enough mass to become spherical in shape can be called as planet, then Pluto could be distinguished from the millions of Kuiper Belt objects. However, during the end of the last century, a few Kuiper Belt objects were discovered that are similar in size and mass to Pluto. It was soon understood that the number of such objects would increase rapidly in the near future, making the number of planets in the solar system very large and ever increasing. In the mean time, in 1995, the first planet outside the solar system and around a Sun-like star was discovered, and subsequently several planets rotating around other stars were discovered. The surprising fact was that all these newly discovered planets are 10–15 times heavier than the largest and the heaviest solar planet, Jupiter. If Pluto too were considered to be a planet, this would have made a huge range of mass among the planets. All these developments in our understanding of planets compelled the International Astronomical Union to set up a committee consisting experts from various fields—physics, chemistry, astronomy, geology, climatology, anthropology, etc.—in order to discuss, debate, and decide a proper definition of planets. After about two years of hectic discussions, the committee proposed that an object could be called a planet if it rotates around a star, does not have its own source of energy, and is sufficiently massive so that it has a spherical shape by its own gravitational force. Obviously, Pluto and Eris fit the bill and were recommended to be planets. Even Ceres and Pallas which were considered as planets long ago and subsequently lost their status as planets once again became planets. However, due to the rapid increase in the number of a large variety of planets outside the solar system, and due to the discovery of another class of objects—Brown Dwarfs that very much resemble Jupiter in nature but several times heavier than Jupiter—the International Astronomical Union considered the definition to be applied only for the solar planets. On 24 August 2006, during the last day of the 26th General Assembly of the Union held at Prague, the resolution was put for discussion and voting. During this time a third criterion was proposed and added to the recommendation of the committee. This condition states that the object should be sufficiently heavy to clear its orbit. Therefore, the amount of mass needed to make the shape of the object spherical by its own gravitational force is not sufficient for it to be considered as a planet. In addition, the orbit of the object should not have any other object or objects, the total mass of which is greater than or comparable to the mass of the main object. Now the mass of Pluto is found to be just 7 % of all the objects rotating around its orbit. So, Pluto does not satisfy the third criteria and hence could not get the same status of the other solar planets. Similarly, Eris and other objects in the Kuiper Belt and Ceres and other objects in the asteroid belt too could not satisfy this condition. Once the resolution regarding the definition of solar planets was passed, the number of planets in the solar system reduced to eight and a separate class was created. This new class is called the dwarf planet which includes Pluto and all other similar objects. So the dwarf planets satisfy the first two criteria but fail to satisfy the third and the new one. As we discussed in the previous chapter, this incident in astronomy is not at all new and so not revolutionary. But it is the result of our enhanced understanding and knowledge on the planets. It is welcome by many, criticized by some, but officially accepted by the astronomical community. The present definition has a caveat: it does not say anything about the maximum mass that a planet can have. Can we tell an object is a planet if it satisfies all the three conditions but 20 times heavier than Jupiter? Can we tell an object is a planet if it satisfies all the three conditions but possibly born in the same way a star is born? In order to avoid such ambiguity, IAU explicitly states that the present definition of planet is applicable to the solar system only. So the debate continues.

Dwarf Planets

Currently there are five recognized dwarf planets. According to their distance from the Sun, they are Ceres, Pluto, Haumea, Makemake, and Eris. Ceres is located at the asteroid belt, between Mars and Jupiter. All the other four dwarf planets are located at the Kuiper Belt. All these five dwarf planets are orbiting the Sun, but they are neither planets nor satellites. However, they are sufficiently massive to have round shape caused by their own gravitational force. The objects that are not massive enough to have spherical shape by their own gravity are called "small solar system bodies." In the future many more dwarf planets will certainly be discovered. In fact, there are as many as 200 objects waiting to be confirmed as dwarf planets.

Pluto

Although Pluto is no more considered to be a planet, it is relevant to know its mass and size as compared to the planets. Pluto was discovered on 18 February 1930 by Clyde W. Tombaugh at the Lowell Observatory. It is one sixth of the size of the Earth and about half of the size of Mercury. Pluto is about six times lighter than the Moon of the Earth. Out of its five satellites discovered till date, Charon, the largest moon of Pluto, is about half the size of Pluto and only eight times lighter than Pluto. That makes both Pluto and Charon orbit around each other (Fig. 3.7). Further, while the Moon is tidally locked with the Earth but the Earth is not, Pluto and Charon both are tidally locked to each other. So the Pluto–Charon system may be more like a

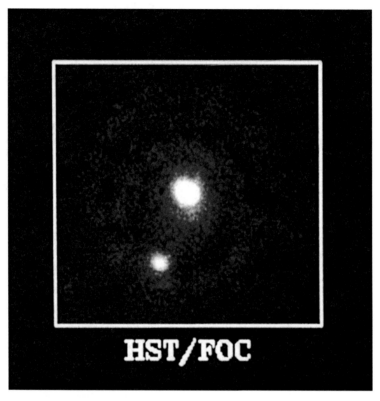

Fig. 3.7 Pluto and its large satellite Charon (Credit: Hubble Space Telescope/NASA)

binary system. At the time of discovery, it was thought that Pluto was as heavy as the Earth. It is worth mentioning here that while Pluto was discovered in 1930, Charon was discovered in 1978 by James Christy. The other four satellites are Nix, Hydra, Styx, and Kerberos. They were discovered only recently. In fact the fourth satellite Kerberos was discovered in 2011 and the fifth one, Styx, was discovered in 2012. Pluto is about a maximum of 49 AU and a minimum of 29.5 AU away from the Sun, and so it is located in the Kuiper Belt. Because of its highly elliptical orbit, Pluto sometimes crosses the orbit of Neptune and hence becomes closer to the Sun than Neptune for about 20 years. It takes Pluto about 248 Earth years to rotate around the Sun. Pluto rotates quite slowly around its own axis—one day of Pluto is equal to about six and a half of the Earth's day. It has a rocky core and a mantle of water ice. Its surface is covered by methane ice. The surface temperature of Pluto is about -230 °C. Its atmosphere consists of about 98 % nitrogen and a little amount of methane and carbon monoxide. When Pluto moves to its farthest position from the Sun, the atmosphere freezes into ice. A thin gaseous atmosphere is formed only when Pluto is at its closest position to the Sun. Nothing much is known about the interior or magnetic field of Pluto. Most probably it does not have a magnetic field. We shall know more about Pluto when the space probe New Horizon will reach near it.

Eris

Eris, once hypothesized as the tenth planet of the solar system, is the heaviest dwarf planet at present. It is 27 % heavier than Pluto and orbiting at a maximum distance of 97.6 AU and a minimum distance of 38 AU from the Sun. We know that Pluto sometimes comes closer than Neptune. Similarly, sometimes Eris becomes closer than Pluto to us. Currently this dwarf planet is at a distance of about 96 AU from the Sun, making it the farthest planetary object in the solar system. Eris takes about 557 years to orbit the Sun.

Eris was detected on 21 October 2003 by Michael Brown of California Institute of Technology. The discovery was announced a bit late, in 5 January 2005. Brown and his team used Palomar Observatory to take the image of this trans-Neptunian object in the Kuiper Belt. Brown nicknamed it Xena after the popular character of the TV serial "Warrior Princess." The official name of this "ex-planet" although was UBU 313. However, after the International Astronomical Union adopted the resolution for the new definition of solar system planets, Eris is reclassified as a dwarf planet, just like Pluto. Subsequently it has got its official name as Eris. Current measurement indicates that Eris is as large as Pluto, about 2,300 km in diameter. Eris has a massive satellite, Dysnomia, which is eight times smaller than Eris and orbiting the dwarf planet in about two weeks.

Similar to Pluto, Eris too has methane ice on its surface. Eris has an albedo of 0.96, second highest among any of the solar planets, dwarf planets, or satellites. The highest albedo of any solar system object is attributed to Enceladus, a moon of Saturn. The surface temperature of Eris varies from $-245\ °C$ to $-220\ °C$ depending on its distance from the Sun. Eris is denser than Pluto, indicating that it too has a rocky core. It has a thin surface of methane and nitrogen ice.

Besides Pluto and Eris, Ceres at the asteroid belt is also quite a large body and is now considered as a dwarf planet. The reader may recall that for a certain period of time, Ceres was considered as a planet. However, the discovery of Eris prompted a debate about the definition of planets as it was realized that many more objects larger than Pluto could be discovered in the Kuiper Belt. IAU's resolution on this aspect was taken in 2006, a few months after the confirmation of the discovery of Eris. So unlike Pluto or Ceres, Eris never achieved an official status of a planet.

Ceres

While Pluto, Eris, Haumea, and Makemake are located at the end of the solar system, in the Kuiper Belt, Ceres is located at the asteroid belt between Mars and Jupiter. Ceres was discovered more than 200 years ago, in 1801. It was the first object discovered in the asteroid belt following the prediction of Titius–Bode empirical formula. Before IAU resolved the definition of planet in 2006, Ceres was considered to be the largest asteroid in the asteroid belt. Ceres is orbiting the

Sun at a mean distance of about 420 million km or 2.8 AU, and it takes about 4.6 years to rotate around the Sun. Ceres has a mean radius as small as about 476 km, and it takes about 9 h to spin around its own axis. So one day of Ceres is only nine hours. Although Ceres is very light, about hundred times lighter than the Moon, it comprises of about one third of the total mass of the asteroid belt. Astronomers believe that Ceres is a failed planet. At the time of the birth of the solar planets, the gravitational perturbation from Jupiter prevented the object from becoming a planet. The mean density of Ceres is about 2 g/cm^3, indicating the rocky nature of the object.

Ceres is quite different than the other objects in the asteroid belt. The mean surface temperature of Ceres is about -105 °C. Ceres may have a very thin atmosphere containing water vapor. There is an indication that Ceres contains large amount of frozen water beneath its surface. Similar to the Earth and Mars, Ceres has a rocky core and icy mantle. Ceres is believed to be geologically inactive. The other relatively large asteroids such as Pallas, Vesta, Juno, etc. are so small that they deviate from spherical shape due to insufficient gravitational force of their own. Therefore, they are not even considered as dwarf planets although once upon a time, they were considered as planets.

The Large Moons of the Solar Planets

Satellites of various planets and dwarf planets show their dominant presence in the solar system. There are hundreds of moons with size ranging from a few kilometers to a few thousand kilometers. The smaller ones like Deimos and Phobos of Mars or Pandora of Saturn or Amalthea of Jupiter are not spherical in shape. They are not sufficiently heavy to become spherical by their own gravitation. On the other hand the larger ones are even bigger and heavier than the dwarf planets. Titan, the satellite of Saturn, even has an atmosphere. On the other hand, Charon and Dysnomia, the satellites of Pluto and Eris, respectively, have a comparable mass to their primaries. These two satellites are so massive that the center of rotation or the barycenter of the system around which the two objects—the satellite and the planet—rotate is outside both the objects. This means Pluto and Eris have not cleared their orbit and so could not satisfy the criteria to become planets. The largest satellite in the solar system is Ganymede, a massive rocky moon of Jupiter. The Moon of the Earth with a radius of 1,738 km is only the fifth largest satellite in the solar system. Ganymede has a spherical shape with radius of about 2,640 km. The third and the fourth largest satellites also belong to Jupiter. They are Callisto with a radius of 2,400 km and Io with a radius of 1,820 km. Europa, the fourth largest satellite of Jupiter, is slightly smaller than the Moon of the Earth. Its radius is about 1,570 km. The second largest satellite is Titan orbiting Saturn. It has a radius of 2,575 km. Triton, the satellite of Neptune, has a radius of 1,353 km. All these seven satellites are larger than the largest dwarf planet Eris which has a radius of about 1,200 km. We shall discuss about some of the remarkable satellites in more detail.

Ganymede

Ganymede is the largest natural satellite in the solar system orbiting the largest planet Jupiter at a distance of about 1 million km from it. Ganymede takes about seven Earth days to complete one orbit around Jupiter. This satellite is even larger than the planet Mercury with a radius of 2,631 km. Ganymede is also the heaviest satellite in the solar system with about double the mass of the Moon. This satellite is tidally locked with the other two heavy satellites of Jupiter—Europa and Io. During one orbit of Ganymede, Europa completes two orbits and Io completes four orbits around Jupiter. This phenomenon is known as orbital resonance. Although Jupiter is a gaseous planet, Ganymede is a rocky object with a small molten iron core and a rocky silicate mantle above the core. The mantle is covered by icy shell. Several craters are found on the surface of Ganymede. Most probably Ganymede has an active geology, e.g., plate tectonics. Dynamic of conducting material inside the satellite gives rise to its own magnetic field which is embedded with the huge magnetic field of Jupiter. Ganymede is the only satellite in the solar system to have its own magnetosphere. The geomagnetic axis is tilted by 176° with respect to the rotation axis. Ganymede possibly has a very thin atmosphere consisting mainly of molecular oxygen which is produced by the dissociation of water ice in its surface. Trace of ozone is also detected. Ganymede has several thick layers of water oceans separated by ice. It may have liquid water just above the mantle.

Titan

Titan or Saturn IV is the only satellite in the solar system to have clouds and an atmosphere. Also, it is the only object other than the Earth wherein matter in liquid form exists. Titan is about half the size of the Earth and almost of the same size of Mars. Titan has a dense atmosphere that extends about 600 km above its rocky surface. The surface temperature of Titan is about -180 °C which freezes water into hard ice and methane into liquid. The atmosphere of Titan contains about 95 % nitrogen and about 5 % methane. Compounds containing carbon, hydrogen, and oxygen are also found in Titan's atmosphere. The liquid methane evaporates to form methane cloud. Titan's surface is covered by sands. Titan does not have a magnetic field or an ozone layer. Therefore, strong ultraviolet rays from the Sun should have destroyed methane as it is the case in Mars. But quite surprisingly, a large amount of methane in liquid as well as in gaseous form exists on Titan. This indicates volcanic activities in Titan which bring the compound from an internal reservoir. Since most of the time Titan is within the magnetosphere of Saturn, it may be possible that Saturn's magnetic field protects Titan from the solar wind. However, the interaction of methane with ultraviolet rays at the uppermost atmosphere produces organic materials such as ethane, hydrogen, cyanide, etc. Titan has rain, wind, etc. that give rise to river and lakes of liquid methane.

Triton

Triton, the largest satellite of Neptune, is most possibly born in the Kuiper Belt and later on captured by the planet. It is about 0.35 million km away from Neptune. Triton is about 2,700 km in size. It has a retrograde orbit which means it orbits the planet in the opposite direction as Neptune spins. Since it orbits backward, it suffers a gravitational torque that makes the satellite lose its orbital rotational energy. As a consequence, Triton is approaching towards Neptune slowly and ultimately it may crush onto the planet.

Triton's spin axis is tilted by 157° with respect to the spin axis of Neptune which in turn is tilted by 30° with respect to the orbital plane of the Sun. Therefore, just like the planet Uranus, this satellite roles along its side and hence the poles are alternatively facing the Sun. Triton has an atmosphere consisting mostly of nitrogen and a little amount of methane. It has a rocky surface with 25 % water ice. It has a high albedo or reflectivity (about 0.8) because of the presence of water ice on the surface. The surface temperature of Triton is 235° below the freezing point of water.

Triton's surface shows the presence of a few craters. Interestingly, Triton is one of the four objects in the solar system to show volcanic activities. It has ice volcanoes erupting liquid nitrogen, dust, and methane. But unlike the volcanoes in Venus or the Earth, Triton's volcanoes are driven by seasonal heating of the Sun.

The Moon

Our Moon, the only natural satellite of the Earth, is our closest celestial object and the second brightest object to us, after the Sun. It rotates around the Earth in 27 days, 7 h, and 43 min with an average speed of 1 km/s. It is about one third the size of the Earth. The Moon is the fifth largest natural satellite in the solar system, and it is the second densest satellite after Jupiter's satellite Io. The mean density of the Moon is 3.3 g/cm^3. It is believed that immediately after the birth, the Earth collided with a very large object—as large as Mars—and the Moon was formed out of the ejected material. Information obtained from the lunar rocks supports this theory. Similar to Mars, the Moon might have a magnetic field at a very early stage, but presently it has no magnetic field. The Moon has no atmosphere either because of its low surface gravity. The surface gravity of the Moon is one sixth to that of the Earth. Therefore, a person weighing 60 kg on the surface of the Earth would have a weight of just ten kilogram on the surface of the Moon. The Moon has several craters named after famous philosophers or scientists such as Copernicus, Tycho, Gagarin, et al.

The reflectivity or albedo of the Moon is about 0.14, much less than that of the Earth. In that sense the Moon is quite a dark object. The mean temperature of the Moon is about −50 °C. It is a perfectly spherical object. The Moon has a crust with a mean thickness of about 70 km. The crust covers the mantle and most possibly there is a core of 350 km radius. However, the Moon is geologically inactive. There is no "Moonquake" or volcanic activities on it. Interestingly, a good amount of water in the form of hydroxyl molecules is detected in the Moon.

The Moon is so near and so heavy that it has a significant influence on the Earth through gravitational force. The tides in rivers, seas, and oceans on the Earth are caused by the Moon. The gravitational attraction of the Moon causes two bulges in the Oceans, one in the direction of the Moon and the other in the opposite direction. Since the Earth rotates much faster than the orbital movement of the Moon around the Earth, these bulges move around the Earth in a day giving rise to two high tides per day. The gravitational force of the Moon is however not along the straight line joining the centers of the Earth and the Moon because of the rotation of the Earth. This produces a torque on the Earth which slows down the Earth's rotation. On the other hand, it accelerates the orbital motion of the Moon because the Earth is much heavier, and so the Earth's gravitational force is much stronger than that of the Moon. Consequently, the Earth's rotation is slowing down by 1.5 s in 100,000 years. At the same time, the orbit of the Moon is getting larger by 3.8 cm per year. Such gravitational interaction also causes the Moon to rotate synchronously around its own axis such that the same side of the Moon always faces the Earth. Before the Moon's rotation was synchronized, it was rotating faster but slowed down by the Earth. On the other hand, the Earth's rotation is being slowed down by the Moon. However, because the Earth's gravitational force is much greater than that of the Moon, the rotation of the Moon slowed down much rapidly. Ultimately, the spin rotation period of the Moon was slowed down to a point that it matches its orbital period so that it always shows the same side to the Earth. Similarly, the Earth's rotation will also eventually slow down to a point at which it will match the Moon's orbital period. However, this will take more than 4 billion years and by that time the Sun will exhaust all its hydrogen, lose hydrostatic equilibrium, and start expanding, and eventually it will engulf the Earth and the Moon. The Moon makes the tilt angle or the obliquity of the Earth's rotation axis stable which helps in making the Earth's climate favorable for us. We shall discuss this in later chapters.

Discovery of Planets Beyond Kuiper Belt: The Boundary Extends

How large is the solar system? Where is its boundary? Is there any edge of the solar system? In 1951, Gerard Kuiper proposed that the outermost boundary of the planet forming proto-planetary disk had a donut-like region wherein the matter was spread so widely that they could not coagulate to form planets during the formation of the solar system. This belt-like region still exists and was discovered in 1992. This is known as the Kuiper Belt. This region extends from the orbit of Neptune at about

30 AU to 55 AU. Kuiper Belt is about 20 times wider and more than 200 times heavier than the asteroid belt that is located between Mars and Jupiter. Any object located beyond the orbit of the Neptune was usually referred to as trans-Neptunian object. After the discovery of the Kuiper Belt, the hundreds and thousands of small rocky and icy objects made of ammonia, methane, and water ices are called Kuiper Belt objects. Asteroids, dwarf planets, and all that orbit the Sun beyond the orbit of Neptune are Kuiper Belt objects. Pluto, Eris, Makemake, and all the dwarf planets are located within this Kuiper Belt and so they too are called Kuiper Belt objects. It is now evident that in the future, many more objects as large as Eris will be discovered in this region. But is the outer edge or the boundary of the Kuiper Belt the edge of our solar system? Can planets or other objects orbiting the Sun exist beyond the Kuiper Belt? Recent discoveries suggest that the solar system is extended far beyond the edge of the Kuiper Belt.

In 1950, a Dutch astronomer Jan Oort suggested that the long-period comets originate from a very distant spherical shell consisting of a cloud of a trillion of small icy bodies or planetesimals. This spherical cloud known as Oort cloud surrounds the solar system. The Oort cloud is believed to contain objects composed of water, ammonia, and methane ices. Due to gravitational perturbation caused by any passing star or activities in the spiral arm of the Milky Way, these comets fall into the inner solar system and appear to us as long-period comets. The comets Hyakutake and Hale–Bopp are two such long-period comets. On the other hand, the short-period comets such as the Halley's Comet come from the Kuiper Belt. In the absence of any direct evidence, Oort cloud still remains hypothetical. However, there are compelling indirect evidences that support the existence of this hypothetical cloud. The Oort cloud can be divided into two regions: a spherical outer region which may extend up to two light years or more than 100,000 AU and a disk-shaped inner region. The outer edge of the inner Oort cloud and the inner edge of the outer Oort cloud are situated at about 1,500 AU away from the Sun. The inner Oort cloud merges with the outer edge of the Kuiper Belt. The outer Oort cloud extends up to a distance where the gravitational influence of the Sun becomes very weak. Therefore, the objects within the inner Oort clouds are more stable under the strong influence of the Sun, but the objects within the outer Oort cloud are easily disturbed by other stars. However, it is still not known exactly how this inner Oort cloud was formed. According to some astronomers, a rogue giant planet was ejected out from its original position and dragged the Kuiper Belt objects out to the inner Oort cloud. Some other astronomers think that an external influence, either by a star or by a group of stars, put the objects into the inner Oort cloud. A third speculation is that the objects within the inner Oort clouds are actually extra-solar objects captured by our Sun. In 2003, a planet-sized object was discovered outside the Kuiper Belt. The estimated distance of this comparatively large object falls at the edge of the inner Oort cloud. Subsequently, one more planetary object was discovered beyond the Kuiper Belt. Thus, the boarder of our neighborhood is extended beyond the Kuiper Belt and to the Oort cloud.

Sedna (2003 VB12)

Sedna was detected on 14 November 2003, but its discovery was reported on 15 March 2004. Its official name is 2003 VB12. By using the Samuel Oschin Telescope at the Palomar Observatory in the United States, Michael Brown, Chadwick Trujillo, and David Rabinowitz discovered this object. The same group of astronomers discovered Eris, the largest known Kuiper Belt object and the largest dwarf planet. Sedna, most possibly a dwarf planet, is presently located at a whooping distance of about 86.3 AU from the Sun. At the time of its discovery in 2003, it was at a distance of about 90 AU from the Sun—almost three times farther than Pluto. The mean distance of Sedna is about 530 AU from the Sun. It rotates in an extreme elliptical orbit. Sedna moves as far as 937 AU from the Sun, but its closest approach to the Sun is 76 AU. Sedna's closest encounter with the Sun will occur after about 62 years from now. It takes about 11,400 years to orbit the Sun. This planet-like object never passes through the Kuiper Belt, but it is certainly far beyond the Kuiper Belt which is extended up to 55 AU from the Sun. The discovery of Sedna implies that the inner edge of the Oort cloud is much closer to us than it was thought before. Its circular orbit might have been perturbed by the gravitational force of a passing star. However, unlike the long-period comets, Sedna was disturbed by stars much closer than previously thought.

The exact size of Sedna is not known. Most likely it has a diameter as small as 1,000 km. So it is smaller than even Charon, Pluto's satellite. It has an albedo similar to that of the Earth. Sedna has no moon and that is why we cannot say how heavy it is. Sedna has about 33 % methane, 10 % nitrogen, 26 % methanol, and a good amount of water ice in its surface. All these materials are in ice form as Sedna is extremely cool with an average surface temperature as low as -260 °C. But nitrogen becomes gas at a very low temperature. The boiling point of nitrogen is about -238 °C and so Sedna sometimes, for a short duration, may form a thin nitrogen atmosphere. Sedna spins around its own axis in about ten hours.

Biden (2012 VP113)

When Pluto was discovered in 1930, it was thought to be a unique object in the Kuiper Belt. Subsequently, several objects including the recently discovered dwarf planets Eris, Makemake, etc. were detected in this region. Similarly, Sedna was also thought to be a unique object in the inner Oort cloud. But very recently a sibling of Sedna in this extended family of the solar system was found which clearly suggests that these are a new group of planetary mass objects. This object, another probable dwarf planet, has even larger orbit and is located beyond Sedna. The distance of its closest approach to the Sun is the longest among all the planets and the dwarf planets discovered till date. 2012 VP113 is popularly known as Biden after the Vice President of the United Sates, Joe Biden. It was detected in 2012, but its discovery was announced in 2014. It was discovered by Scott S. Sheppard and Chadwick Trujillo. This inner Oort cloud object was discovered at Cerro Tololo Inter-American Observatory. The closest distance of Biden (known as perihelion) to the Sun is about 80 AU, slightly larger than that of Sedna. At present this object is located at a whooping distance of 83 AU or about 12.4 billion km from the Sun. Biden is about half the size of Sedna. It should be emphasized here that the International Astronomical Union has not yet recognized Sedna and Biden as dwarf planets. They may not also be called as Plutoids and certainly not as Kuiper Belt objects. More information about them, specially an accurate estimation of their mass, is needed to decide their class. Most likely, they will be categorized by a new class when many more similar objects in the inner Oort cloud will be discovered.

After the discovery of Sedna and Biden, it is realized that a large number of small icy objects must be there in the inner Oort cloud. In fact many of them, smaller than Sedna or Biden, are already detected but yet to be confirmed. Interestingly, the natures of the orbit of both Sedna and Biden are the same, and their dynamics imply that the orbits of both these tentative dwarf planets are disturbed by some unknown massive object in the same region. This suggests the presence of an Earth-sized planet or even a heavier planet at a distance of a few hundred AU from the Sun. Quite a few efforts have been made to detect this unknown Earth or Super-Earth, but it still remains elusive.

Comets, Asteroids, and Meteoroids

This chapter will remain incomplete if we do not talk about the objects much smaller than the planets, satellites, and dwarf planets. These are known as minor solar system objects. There are millions of such objects in the solar system. While comets and asteroids orbit the Sun, meteoroids are small stray objects not larger than 10 m in size. When a meteoroid enters the Earth's atmosphere and burns up as it passes through the atmosphere, it is called a meteor. We often refer the meteors as "falling" or "shooting" stars. If a meteoroid is sufficiently large such that it survives while passing through the atmosphere and subsequently hits the Earth's surface, it is called a meteorite.

Comets are usually made of ice, dust, and other materials such as carbon dioxide, methane, ammonia, etc., while asteroids are made of rocks and metals. Most of the asteroids orbit the Sun in the asteroid belt between Mars and Jupiter. But the short-period comets originate in the Kuiper Belt beyond the orbit of Neptune and the long-period comets originate in the hypothetical Oort cloud, several billion kilometers away from the Sun. An asteroid that approaches the Earth at a distance of less than 1.3 AU is known as Near Earth Object. This happens due to gravitational disturbance in the otherwise circular orbit of the asteroid. Asteroids appear in different sizes and shapes. They can be as large as a thousand kilometers and as small as a few meters. Both asteroids and comets are the leftovers of the formation of the solar system, around four and a half billion years ago. Asteroids could not accumulate enough mass to become planetary objects. Some of the largest asteroids in the asteroid belt are Vesta, Juno, Pallas, etc. Once upon a time they were considered as planets. Comets develop a tail when they approach the Sun, and the tail is directed opposite to the Sun. The heat from the Sun melts and vaporizes the ice and other materials in a comet. The vapor is expelled and trails behind the rocky head known as the nucleus of the comet. The size of the nucleus varies from 1 to 50 km. The nucleus is surrounded by an envelope of haze or cloud called the coma. This trailing vapor of gas and dust forms the tail and extends a million of kilometers from the coma. The strong solar wind blows the tail opposite to the Sun. The tail is illuminated by sunlight and thus it glows and becomes visible in the night sky. Unlike the circular orbits of the asteroids, comets orbit in highly elliptical orbit. Some of the famous and remarkable comets are comet Halley discovered by Edmund Halley in 1531, comet Hale–Bopp discovered by Alan Hale and Thomas Bopp in 1995, comet Swift–Tuttle discovered by Lewis Swift and Horace Tuttle in 1862, comet Hyakutake discovered by Yuji Hyakutake 1996, comet Shoemaker–Levy discovered by Carolyn and Eugene Shoemaker and David Levy in 1993, etc. Comet Shoemaker–Levy collided with Jupiter between 16 July and 22 July 1994. On the other hand meteoroids are small debris of broken objects that rotate at various speeds in a variety of orbits around the Sun. Every year about a hundred of small meteorites hit the Earth. Large meteorites create craters on the surface of the Earth. Some of the meteorites are quite heavy, a few tons in weight. The Hoba meteorite weighs about 60 tons and it is the largest single meteorite known. The Willamette meteorite, the largest known single meteorite found in the United States, is about 15 tons in weight. Comets, asteroids, and meteoroids are carriers of many compounds that were not present in the early Earth.

Can There Be Life in Other Solar Planets?

Once we learn about the planets, dwarf planets, satellites of the planets, etc., we naturally become curious to know if any of these objects or members of the solar family harbor life. We would like to know if life exists within our immediate neighborhood. But this needs an explanation about what we mean by life. While a detail discussion on this issue is postponed to later chapters, we can pose the question in a different way. Can any form of terrestrial life, micro or macro, originate and survive on other solar planets? Or simply can any living species on the Earth survive naturally if taken to any other solar planets? We know that only the inner four planets Mercury, Venus, the Earth, and Mars as well as a few of the moons around both the inner and the outer gaseous planets have solid surface. Mercury is too hot and it has no atmosphere. The atmosphere of Venus mainly consists of carbon dioxide. Recently a thin layer of ozone is reported at the night side of Venus. However, lack of molecular oxygen and liquid water makes it impossible for life to survive in Venus. The high surface temperature that can melt even some metals such as lead, tin, or even zinc and absence of liquid water ruled out life on Venus. The Moon has no atmosphere, neither does it have a magnetic field. The water molecules are dissociated into hydroxyl by the ultraviolet rays and charged particles from the Sun. The planet Mars and the satellite Titan could have a supporting climate. Among all the satellites, only Titan has a dense atmosphere. But Titan is too cold for life to survive. It is so cold that water exists in hard solid form and methane exists in liquid form. There is no adequate amount of oxygen in the atmosphere of Titan. However, if the Sun becomes much brighter in the future, Titan may achieve an atmosphere favorable for the survival of life.

Mars has many similarities with the Earth and hence Mars may be considered as the next or the only planet beyond the Earth to exhibit some possibility of harboring life. Mars has a solid surface and the total dry land is almost equal to that on the Earth. It has almost the same duration of day and night that the Earth has. The tilt in its rotation axis at present is almost the same to the tilt of the Earth's rotation axis, and hence, Mars has seasonal variation. It has an atmosphere. However, the important conditions necessary to protect life are missing in Mars. The atmosphere of Mars mainly consists of carbon dioxide and there is no free oxygen. There is not much evidence for the presence of liquid water. Further it has no magnetic field to protect life against the energetic charged particles of the solar wind. Carbon dioxide gets easily converted into the poisonous carbon mono-oxide by the intense ultraviolet rays of the Sun which penetrate deep inside Mars atmosphere. Mars does not have geological activities. In the absence of volcanic activities or plate tectonic, the minerals in Mars are not recycled, and so the rocks in Mars are as old as about three and a half billion years. It is believed that Mars has a large reservoir of water beneath its surface. But there is no way for this water to come out to the surface. Mars has an elliptical orbit and so it warms up when it comes nearer to the Sun yielding the temperature suitable for water to exist in liquid state. However, when it moves farther to the Sun, it becomes extremely cool. So it is very much unlikely that any living species even microscopic one can survive on Mars. Still some people believe that life could have formed on Mars. If so, life at least in the form of microbial that can survive extremely hostile conditions may exists inside the rocks of the planet. Detection of methane can indicate the presence of such life. However, none of the various probes to Mars that orbited the planet very closely or even landed on the surface could detect any bio-signature. Nevertheless, the possibility of the existence of nascent life on Mars is not completely ruled out. Even the existence of life in early Mars remains debatable.

Chapter 4
Brown Dwarfs: The Missing Link Between Stars and Planets

> *As one great furnace flamed, yet from those flames*
> *No light, but rather darkness visible*
>
> – John Milton
> (*In* Paradise Lost)

Brown Dwarfs: The Missing Link

It is well known that the lightest stars are about hundred times heavier than Jupiter. Before the discovery of planets outside the solar system, Jupiter was known to be the heaviest object that was not a star. This huge gap in the hierarchy of mass was quite a puzzle. So, quite naturally, astronomers were curious to know if there was any celestial object lighter than the lightest star but heavier than Jupiter. Is there a missing link between the stars and the planets? In the year 1963, Shiv S. Kumar, an Indian origin American astronomer, theoretically predicted that such objects might exist. It was also realized that such kind of objects, if existed, should be extremely faint. Kumar named them "Black Dwarfs." But the name "Black Dwarf" was already assigned to the end product of a kind of objects called White Dwarfs. White Dwarfs are the final state of a Sun-like star. After burning for several billion years, when the nuclear fuel, e.g., hydrogen, gets depleted at the core of a star, helium starts burning at the central region or the core and hydrogen starts burning outside the core, at the shell of the star. This generates so much heat energy that gravitation cannot hold the star and so the star starts expanding. When our Sun will undergo such a stage, it will expand so far out that even the Earth will be engulfed by it. Such an expanded star is known as Red Giant star. All stars, irrespective of how heavy they are, pass through this stage at the end of their life. Once helium gets depleted and the nuclear burning stops, radiation force becomes weak and the core becomes heavier. As a consequence the star starts contracting due to the gravitational pull of the matter at the core. For a star like our Sun, the contraction stops at a state called White Dwarf. A White Dwarf is as small as the Earth but as heavy as the

Sun. If the mass of the stellar core is more than 1.4 times the mass of the Sun, the celebrated Chandrasekhar limit, the contraction stops at a form known as neutron star which has a radius of just 10 km—smaller than the size of even a town or a city. On the other hand, if the star is eight to ten times heavier than the Sun, nothing can stop the gravitational collapse and the object becomes the so called "black hole." White Dwarfs emit light through the release of gravitational potential energy generated due to contraction. When a White Dwarf cools down completely and becomes invisible, it is called Black Dwarf. However, it takes such a long time for a star to become Black Dwarf that we do not yet find any of them. Actually, the time taken for a star to become Black Dwarf is longer than the present age of the universe. So, in 1975, astronomer Jill Tarter renamed the objects that were theoretically predicted by Kumar as Brown Dwarfs. It was predicted that these objects would be bright in infrared light which is beyond the visible range of light. It would emit radiation at a wavelength longer than that of red light. So, in order to make an analogy with visible colors, the name Brown Dwarf was given. However, such kind of objects was elusive for more than three decades. Quite a few astronomers claimed to have discovered it, but careful analysis refuted their claims and most of the time a low-temperature M-dwarf star was mistaken as a Brown Dwarf. Finally, in the year 1995, the discovery of a genuine and unimpeachable Brown Dwarf orbiting around a Red Dwarf star Gliese 229 was confirmed by a team of astronomers from Caltech and Johns Hopkins University. This object is located at about 19 light years away from us. In the same year, the discovery of an isolated Brown Dwarf in the Pleiades open cluster was also reported by a group of Spanish astronomers. This object known as Teide 1 after the Teide Observatory is located at a distance of about 400 light years from us. Incidentally, in the same year, the discovery of the first planets outside the solar system was also confirmed.

Now, what are these Brown Dwarfs? Are they stars or planets? The answer is neither. So let us first discuss why a Brown Dwarf cannot be considered as a star.

Brown Dwarfs: The Failed Stars

A star is formed by the gravitational collapse of a massive molecular cloud or nebula. When the collapse is halted, the resulting object is called a star only if it can produce energy continuously at its core. The source of this energy is nuclear. A star produces energy by fusing two atoms at its core. We have heard about hydrogen bomb that produces enormous amount of energy. This energy is produced by the fusion of two hydrogen atoms. Inside the Sun or a Sun-like star, a million of hydrogen bombs are producing enormous amount of energy continuously. However, stellar nuclear fusion involves a chain process. An atom consists of a nucleus around which negatively charged particles called electrons revolve. The atomic nucleus contains two kinds of particles—protons and neutrons. Neutrons are electrically neutral particles. The nucleus of an ordinary hydrogen atom however does not have a neutron. By the fusion of atoms, we mean the fusion of two nuclei. A proton has a positive electric charge and we know that the same electric charges repel each other. Therefore, two protons under normal condition cannot come near each other. When the star-forming cloud collapses, the core or the central region of the cloud becomes extremely hot. As a result, all the protons inside the core start moving very fast. The high pressure as well as the thermal energy makes two protons to overcome their electromagnetic repulsive force. When the temperature of the core reaches about three million degree Celsius, two protons from two hydrogen atoms fuse with each other producing the nucleus of a deuterium atom. A deuterium atom has one proton and one neutron in its nucleus. In this process a positively charged electron called positron emerges out along with energy. Positron is actually an anti-electron. Thus, although two positively charged protons combine, the nucleus of deuterium has the same charge to that of a single proton. Deuterium is a stable isotope of hydrogen atom and is also known as heavy hydrogen. This deuterium almost immediately combines with another proton to yield an isotope of helium. This newly produced helium isotope has two protons and one neutron in its nucleus. This process is commonly known as deuterium burning process. However, the stellar nuclear burning or hydrogen burning can continue and becomes a continuous source of energy only if two such helium isotopes can fuse to produce another isotope of helium with two protons and two neutrons in its nucleus. But such helium burning process needs a temperature much higher than that needed by the deuterium burning process.

The necessary amount of heat that initiates deuterium burning process is produced only if the core of the collapsing cloud or the protostar is at least 13 times heavier than Jupiter. But helium burning process needs much higher temperature—about 6–8 million degree Celsius. This extremely high temperature at the core can be achieved if the object is at least about 80 times heavier than Jupiter. On the other hand, nuclear burning can continue only if the core can initiate the helium burning process. Therefore, the protostar can become a star only if it is at least 80 times heavier than Jupiter. In other words, an object ignites deuterium if it is more than 13 times heavier than Jupiter, and it can ignite helium to become a normal star if it is more than 80 times heavier than Jupiter. Brown Dwarfs lie between these two limiting masses. Thus, a Brown Dwarf initiates nuclear burning process but fails to sustain it, and hence, it does not have a continuous source of energy at its core. In that sense, a Brown Dwarf is a failed star. So Brown Dwarfs are born in the same way a star is born, but they fail to become normal stars due to insufficient mass. They are at least 13 times heavier than Jupiter. On the other hand, a planet cannot even ignite deuterium, and hence, a planet must be lighter than the lightest Brown Dwarf. Therefore, Brown Dwarfs link the least massive stars and the most massive planets. No planet can have a mass more than 13 times the mass of Jupiter.

Climate of Brown Dwarfs

Since Brown Dwarfs are not sufficiently massive to become a star, they belong to a class known as substellar mass objects. Although the existence of Brown Dwarf was theoretically predicted during the 1960s of the last century, it took quite a few decades to discover them. The main reason behind the delay in discovering them is that these objects are extremely faint. Their light originates from the conversion of gravitational potential energy into heat energy during the collapse. They are brighter in the infrared than in the visible light. The discovery of Brown Dwarfs was claimed several times by different astronomical groups. But in most of the cases, the object discovered was found to be a faint star—M-dwarf. Finally, in 1995, a faint small object orbiting a Red Dwarf star Gliese 229 was reported to be the first confirmed Brown Dwarf. It was found to be fainter than the faintest stars, its mass was estimated to be less than 80 times of Jupiter mass, and its temperature was also found to be less than the temperature of the coolest stars. But the presence of methane in its atmosphere has vindicated the claim that Gliese 229b (the primary Red Dwarf star is known as Gliese 229a) is indeed a Brown Dwarf—the elusive missing link between stars and planets. Methane molecules cannot exist under the high temperature of a normal star. On the other hand, signature of primordial lithium vindicated that Teide 1 was also a substellar mass object or Brown Dwarf because lithium is destroyed inside a normal star within 100 million years. Till date a large number of Brown Dwarfs are discovered.

Brown Dwarfs are as large as Jupiter and about ten times smaller than the Sun. The temperature at their photosphere, i.e., the region from which light comes out to the observer, ranges between 1,900 °C and −50 °C. In Chap. 1, we have discussed about the classification of stars on the basis of their atmospheric composition as implied by their spectra. Before the discovery of Brown Dwarfs, the stars were divided into seven classes. However, the spectra of Brown Dwarfs differ so much from that of the coolest stars known as M-dwarfs that three more classes were introduced into the stellar spectral classification—L-, T-, and Y-dwarfs. The photospheric temperature of L-dwarfs ranges from 1,900 °C to 1,100 °C. Not all L-dwarfs are Brown Dwarfs or substellar mass objects. Only those L-dwarfs that show clear signature of primordial lithium in their spectra are Brown Dwarfs. Primordial lithium was produced during the Big Bang nucleosynthesis. Primordial lithium gets destroyed inside a star that has ignited helium. The Brown Dwarfs whose photospheric temperature lies between 1,100 °C and 400 °C are known as T-dwarfs or Methane-Dwarfs. The coolest Brown Dwarfs whose photospheric temperature is below 400 °C are called Y-dwarfs. All T- and Y-dwarfs are Brown Dwarfs or substellar mass objects. Y-dwarfs, the coolest and the faintest Brown Dwarfs, have been discovered very recently, after more than a decade of the discovery of the first Brown Dwarf Gliese 229b. In fact the fourth closest stellar object to our Sun is a Y-dwarf designated as WISE J0855-07144. WISE stands for Wide-field Infrared Survey Explorer. It was a space telescope of NASA. This object located at about 7.2 light years from the Earth is as cool as the North Pole of the Earth. The estimated temperature of WISE J0855-07144 ranges between

−48 °C and −13 °C. Interestingly, the third closest stellar object to the Sun is a system of two Brown Dwarfs called Luhman 16A and Luhman 16B after their discoverer Kevin Luhman. Luhman 16A is an L-dwarf having a photospheric temperature of about 1,100 °C, while Luhman 16B is a T-dwarf with a photospheric temperature about 937 °C. They orbit each other at a distance of 3 AU and take about 25 years to complete an orbit. This binary system is about 6.5 light years away from the Sun. Luhman 16A and Luhman 16B are the closest Brown Dwarfs to us. It is suspected that this binary Brown Dwarf system has a planetary mass object as a third component.

Most of the matter inside a Brown Dwarf is in boiling state. Above this boiling or convective region, there exists a thin atmosphere. The atmosphere of a Brown Dwarf is sufficiently cool for different molecules to form out of different atoms. Brown Dwarfs are very dense as well. Their surface gravity is at least ten times and at most a hundred times greater than the surface gravity of Jupiter. The surface gravity of an object is directly proportional to its mass and inversely proportional to the square of its radius. The weight of a body on any celestial object is determined by the product of its mass and the surface gravity of the object. Since the mass of any Brown Dwarf is at least 13 times higher than the mass of Jupiter but they are as small as Jupiter in size, the density of a Brown Dwarf is much higher than the density of Jupiter. Depending on their atmospheric temperature, molecules such as methane, ammonia, carbon dioxide, carbon monoxide, and water vapor are formed in the atmosphere of Brown Dwarfs. The appropriate combination of temperature, pressure, and surface gravity allows the gaseous compounds of silicon to condensate and to form dust cloud in the visible atmosphere of L-dwarfs, the relatively hotter Brown Dwarfs. L-dwarfs are so hot that water cloud would evaporate from their atmosphere. Clouds consisting of forsterite, gehlenite, silicon oxide, or sand, etc. are detected in the atmosphere of L-dwarfs. Therefore, L-dwarfs have a hot and cloudy atmosphere (Fig. 4.1). As the objects become older and hence cooler and transit from L- to T-dwarfs, the dust cloud rains out, i.e., the dust grains are precipitated below the visible region. Molten sands and other compounds of silicate rain when the atmosphere of an L-dwarf cools down. Therefore, the atmosphere of T-dwarfs is cloud-free. However, Y-dwarfs are sufficiently cool to have ammonia and water cloud in their atmosphere.

Many Brown Dwarfs are found to be rotating very fast along their own axis, and therefore, it is inferred that they are not exactly spherical but oblate just like the giant solar planets. Although it is expected that they should have weak magnetic field, not much is known about it. A few Brown Dwarfs are found to have planets around them.

As we shall see in the subsequent chapters, the temperature range of Brown Dwarfs overlaps with many hot extra-solar planets that are orbiting very close to their parent stars. In fact the surface temperature of many Y-dwarfs is much less than that of the Earth. Hence, the atmosphere of Brown Dwarfs should resemble that of many extra-solar planets. Since Brown Dwarfs have their own radiation, they are not hidden by the intense starlight, and they are sufficiently bright, they can easily be observed through moderately large telescopes. A good understanding of

Fig. 4.1 Image of an L Brown Dwarf 2MASS J0036+1821 (Image obtained by using the 2 m Himalayan Chandra Telescope, Indian Astronomical Observatory, Indian Institute of Astrophysics)

their atmospheric properties would provide important insight on the climate of various extra-solar planets which cannot be seen directly.

Interestingly, a companion Brown Dwarf in a binary stellar system, i.e., a Brown Dwarf orbiting around a normal star such as Gliese 229b, satisfies all the criteria for the definition of planets given by the International Astronomical Union. Brown Dwarfs are in hydrostatic equilibrium. They don't have their own source for producing energy continuously and they have certainly cleared their orbit. IAU has neither fixed any upper limit on the mass of planets nor does it say anything about deuterium burning. On the other hand, it is not possible to infer the birth history of a substellar mass object, i.e., whether it was born like a star or like a planet. Therefore, all the Brown Dwarfs rotating around a star may qualify to be planets. In order to avoid this contradiction, IAU has specifically mentioned that the current definition of planets is applied only to the solar system. It still remains ambiguous to define a border between a Brown Dwarf and a giant planet.

There is one more problem. Stars and Brown Dwarfs are born from the core collapse of interstellar clouds, while planets are born out of the proto-planetary disk. Now detailed numerical calculations indicate that objects having mass as small as twice the mass of Jupiter can be formed out of the core collapse mechanism. Of course they cannot ignite deuterium at their core. What should be the status of such an object? Should we call them planets or should we term them sub-Brown Dwarfs? Since there is no way to know how such an object was born, by core collapse or through proto-planetary disk, it's very ambiguous to characterize any planetlike object which is a few times heavier than Jupiter. An isolated planetary mass object or rogue planet which does not orbit around any star may either be a sub-Brown Dwarf or a planet detached from its parent star. This is an unresolved problem. For the time being, astronomers in general consider the deuterium burning mass limit as the boundary between planets and Brown Dwarfs. This means, any object which has mass less than 13 times the mass of Jupiter is considered as a planet. So nature confuses us by creating such a variety of objects in the universe.

Chapter 5
Discovery of Extra-Solar Planets

> *I do not know what I may appear to the world, but to myself
> I seem to have been only like a boy playing on the sea-shore,
> and diverting myself in now and then finding a smoother
> pebble or a prettier shell than ordinary, whilst the great
> ocean of truth lay all undiscovered before me.*
>
> —Sir Isaac Newton
> *(In Memoire of the life, Writings, and Discoveries of Sir
> Isaac Newton by Sir David Brewster)*

51 Pegasi b: The First Confirmed Customer

As soon as the Copernican concept of heliocentric world was established, Giordano Bruno, a sixteenth-century Italian philosopher, and two centuries later Sir Isaac Newton predicted that many of the stars fixed in the sky should also have planet and planetary system around them. During the nineteenth and the twentieth centuries, quite a few astronomers claimed to have discovered planets around other stars. Most of these claims were based on indirect evidences such as anomalies in the orbital motions of binary stars or variation in the brightness of apparently isolated stars. However, none of those claims were confirmed and subsequently they were rejected by the astronomical community. Among all such reports, the most significant one was made in 1855 by Captain William Stephen Jacob, director (1849–1858) of the then East India Observatory at Madras in India (this observatory was the origin of the Indian Institute of Astrophysics, Bangalore). From the study of the orbital anomalies of the binary star 70 Ophiuchi, Captain Jacob inferred the presence of a third object in this system with a mass comparable to the typical mass of a planet. He suggested that the orbital anomalies of the stars in this system could be explained if there was a planet orbiting the binary stars. His work entitled "On Certain Anomalies presented by the Binary Star 70 Ophiuchi" was published in the Monthly Notices of the Royal Astronomical Society (Vol. 15, pages 228–231). This suggestion was supported by Thomas Jefferson Jackson See of the University of Chicago in 1896. Using Leander McCormick Observatory of University of Virginia, See even deduced the orbital period of the planet as 36 years and his work was published by the Astronomical Journal (Vol. 16, pages 17–23). However, a detailed study on the stability of such a system of triple objects ruled out the possibility of the presence of a planet. Later on it is found that the calculations by both Jacob and See were erroneous. Historically, Captain Jacob's report is considered to be the first "false alarm" for extra-solar planet. However, the issue whether 70 Ophiuchi indeed has a planet or not is still unresolved. In recent years, astronomers from McDonald Observatory of US estimated an upper limit on the mass of the planet or a planetary system around 70 Ophiuchi. Similarly, in 1950, the discovery of a planet around the Barnard's star, the second closest star to the Sun, was claimed, and in 1991, a planet around the pulsar PSR 1829-10 was suggested. Note that pulsars are fast rotating neutron star. Strictly speaking they are not normal stars but the end product of normal stars that were much heavier than the Sun. However, all these claims were rejected after detailed and systematic follow-up observations. In astronomy, a discovery by individual astronomer or a group of astronomers gets recognition only after it is verified and confirmed by other astronomers. However, sometimes confirmation is reported quickly and sometimes it takes longer time for verification. Therefore, the date or year of a discovery and the date or year of confirmation of the discovery both are significant in the history of astronomy.

In 1988, Bruce Campbell, G. A. H. Walker and S. Yang claimed the discovery of a planet around Gamma Cephei. However, their claim remained controversial and unconfirmed for a long 15 years. In 2003 when improved observational facilities were available, this discovery was confirmed. In that sense the planet around the star Gamma Cephei could be considered as the first discovery of an extra-solar planet. A few years later, in 1992, Aleksander Wolszczan and Dale Frail announced the discovery of two planets around the pulsar PSR 1257+12. Their claim generated debates, discussions, and doubts. But eventually it is confirmed that the pulsar indeed has three planets orbiting it.

The first confirmed extra-solar planet around a Sun-like star was discovered in 1995 by Michel Mayor (Fig. 5.1) and Didier Queloz of the University of Geneva. Using a special instrument attached with a telescope at Observatoire de Haute-Provence, they discovered a planet around a Sun-like star 51 Pegasi located about 50 light years away from us. Within a short period, this discovery was confirmed by Geoffrey W. Marcy and R. Paul Butler of the University of California at Berkeley. Marcy and Butler used their instrument at Lick Observatory. The discovery of the planetary system around the pulsar PSR 1257+12 was also confirmed subsequently but only after the discovery and confirmation of the planet in 51 Pegasi. Hence, the planet 51 Pegasi b is considered to be the first confirmed planet outside the solar system. Unofficially this planet is called as Bellerophon. The planets around PSR 1257+12, however, may be considered as the first confirmed planetary system or

Difficult to Detect Extra-Solar Planets Directly

Fig. 5.1 Michel Mayor, the discoverer of the first confirmed extra-solar planet around a normal star (Photograph by the author)

system of multiple planets discovered outside the solar system. The first planetary system around a Sun-like star Upsilon Andromeda was discovered in 1999.

Difficult to Detect Extra-Solar Planets Directly

Since the discovery of 51 Pegasi b, a large number of planets and planetary systems outside the solar system have been discovered within a short span of time. But almost all of these planets are discovered by indirect methods. Why is it difficult to detect planets around a distant star directly? There are two main reasons. Firstly, the planets are extremely faint as they do not have their own source of light. They only reflect

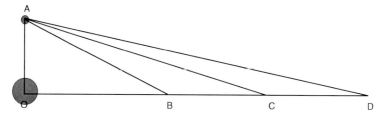

Fig. 5.2 Angular separation of the planet A and the parent star O from an observer at different positions B, C, and D (Illustration by the author)

lights from their parent star. Even a planet as large as Jupiter, orbiting around a Sun-like star at a distance of 1 AU (i.e., at the distance between the Earth and the Sun) would be a billion times fainter than the star in the visible light. In the infrared, the planets are usually brighter but about a million times fainter than the star. Therefore, it's difficult to detect them, especially when their positions are not fixed but alter rapidly around their parent stars. However, a sufficiently large space telescope can do the job even if they are extremely faint. The main problem in detecting a planet around a distant star is the apparent visual separation between the bright star and the faint planet. It is extremely difficult, many a times impossible to separate a planet from the bright star. This is because of the fact that, for a fixed orbital distance between the star and the planet, the angular separation between them is extremely small to an observer at the Earth. It can be realized from Fig. 5.2. The angle subtended by the planet, the observer at the Earth, and the star reduces as the distance between the star and the Earth increases. The angular separation is proportional to the ratio between the orbital distance of the planet from the star (AO) and the distance of the star from the observer (OB). For example, suppose there is a planet orbiting its parent star at a distance of about 0.1 AU or less. If the star is a few light years away from the Earth, the angular separation between it and the planet would be so small that the planet cannot be resolved from the star by any instrument however large or powerful the telescope is. So due to the limitation of instruments, planets orbiting around a distant star cannot be detected directly. However, a planet orbiting a few AU away from its parent star may be detected directly by blocking the bright light of the star. Very recently, a few young and bright planets far away from their host stars have been directly imaged. If the planet is far away from its parent star, we may see them by blocking the starlight, but if the planet is near to the star then we can only feel them. So the question is how these dim, small planets orbiting very close to their bright and large parent stars far away from the Earth can be detected.

Detection Methods

Astronomers use various methods to discover indirectly planets around stars that are far away from the solar system. However, most of the planets are discovered by using two methods, e.g., the Doppler or Radial Velocity method and the Transit method. Other methods include gravitational micro-lens, timing, imaging, etc. Initially a single planet around a star is discovered, but by follow-up studies, many planets around the same star are detected. Thus, many multiple planetary systems around various kinds of stars have already been discovered. These are the worlds beyond our own. Let us now briefly describe some of these planet detecting methods presently employed by the astronomers.

The Radial Velocity Method

In order to discover the first confirmed extra-solar planet around 51 Pegasi, Mayor and Queloz used the method known as Doppler method or the Radial Velocity method. A large number of the extra-solar planets were detected by using this method until the space telescope Kepler was functional. Therefore, the Radial Velocity method has been a very popular method among the extra-solar planet hunters. What is this Radial Velocity method? In 1842, the Austrian physicist Christian Andreas Doppler noted that the pitch of sound changes if the source of the sound is in motion. Doppler postulated that since the pitch of sound from a moving source alters for a stationary listener, the color of the light from binary stars or a moving star should also vary. This was subsequently confirmed by experiment and the phenomenon is known as Doppler Effect.

Natural light, be it from a star or a fluorescent lamp, is a mixture of several electromagnetic waves. Color is a specific sensation to human eyes. Each wave provides a sense of color to human eyes. Each color a normal human eye senses is characterized by a particular wavelength. For example, red light has a different wavelength than green or blue light. If the length of a wave that characterizes red light is reduced, it will be converted into green or blue light to human eyes. This means the wavelength of red light is longer than the wavelength of yellow light which is longer than that of green light which in turn is longer than that of blue light. In the visible range, blue light has the shortest wavelength, while red light has the longest one. The Doppler Effect of light is illustrated in Fig. 5.3. According to this effect, if a source of light (in this case the light from a star) moves away from or towards a stationary observer, the wavelength associated with the light that appears to the observer changes. If the source of light moves away from the observer, the wavelength increases, and if it moves towards the observer, the wavelength reduces (see the white lines above the peak of each wave in Fig. 5.3). The shift in wavelength towards longer wavelength is called red shift as the longest wavelength in the visible region belongs to red light. The shift towards shorter wavelength is called blue shift. In red shift, the energy of light reduces while in blue shift, it increases. The energy is inversely proportional to the wavelength of light.

Now, in a binary system, two objects rotate around each other. The heavier one is called the primary component, and the lighter one is called the secondary component of the system. For a planet–star system, the star is the primary component and the planet is the secondary component of the two-body system. However, none of the two objects rotate around the center of the other object. But the two objects

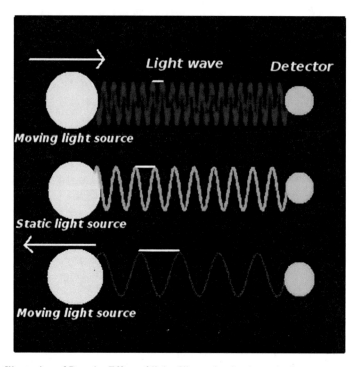

Fig. 5.3 Illustration of Doppler Effect of light (Illustration by the author)

rotate around a common center known as the center of mass of the system or barycenter. The center of mass can be imagined as a balancing point between two objects, and it is located at the line joining the centers of the two spherical objects. It may lie either inside one of the objects or outside both of the objects. If the objects have the same mass, then the center of mass or the barycenter is located exactly at the middle of the line joining the centers of them. Now, if the mass of one of the objects is increased, the center of mass will shift towards the center of this heavier object. Thus, if the mass of the secondary is very less as compared to the mass of the primary, the center of mass around which both the objects rotate will almost be the center of the primary object. Now, for a planet–star binary system, if the planet is very near to the star and is as heavy as Jupiter, the center of mass of the system should be inside the star but not exactly at the center of it. Both the star and the planet should rotate around this center of mass or barycenter. As a result, the star will wobble as illustrated in Fig. 5.4. The velocity at which the star moves towards or away from the observer is called the radial velocity of the star. This radial velocity of the star results into Doppler Effect of light. When the planet moves towards the observer, the star also moves a little towards the observer and hence the starlight is blue shifted. Similarly, when the planet moves away from the observer, the starlight is red shifted. Light from a star received by a telescope is passed through a special instrument called a spectrograph which splits the white light into

Fig. 5.4 Schematic diagram for Radial Velocity method. The position of a dark line in the spectrum corresponds to the position of the planet around the star (Illustration by the author)

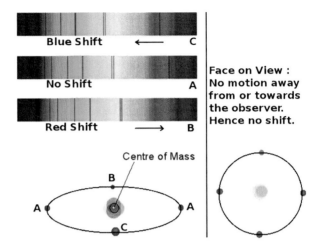

different colors characterized by different wavelengths. The spectrum of the star obtained by this instrument shows several vertical dark lines known as absorption lines (see Fig. 5.4) These absorption lines are formed due to the absorption of light by different elements in the atmosphere of the star. The displacement or shift in the position of these dark lines indicates the Doppler Effect due to the radial velocity of the star. Usually the absorption line due to iodine is monitored by passing the starlight through a cell containing iodine. This enables to detect and measure the displacement of the stellar absorption lines with respect to the static absorption lines of iodine. The shift in the iodine absorption lines would indicate the movement of the star due to the presence of the planet or a secondary component. The amount of shift provides the amount of radial velocity of the star. The amount of radial velocity gives the minimum mass of the planet provided the mass of the star is known. The time taken by an absorption line to move from an initial position, say A to B and then back to A, to C, and return to the same initial position A (as shown in Fig. 5.4), is the orbital period of the planet.

However, if the system is observed face on, the movement of the star will be lateral—neither away from the observer nor towards the observer—and so Doppler Effect will not be detected. The angle between the plane containing the line joining the planet and the star and the plane perpendicular to the line of sight is called the inclination angle of the planet with respect to the observer. When the plane perpendicular to the line of sight is parallel to the plane containing the line joining the planet and the star, the system is viewed face on. So face on view means the inclination angle is zero. On the other hand, if the inclination angle is 90°, the system is viewed edge on. The Doppler shift of a star is the maximum when the system is viewed edge on, i.e., when the inclination angle is 90°. It reduces with the decrease in the inclination angle and is zero when the inclination angle is zero. If the planet is very small, the center of mass of the system would be very close to the center of the star. In that case, the radial velocity of the star should be so small that the shift in the wavelength can hardly be detected in the spectrum. Planet Earth exerts a radial velocity of only 9 cm/s to the Sun, whereas Jupiter exerts a radial velocity of 12.7 m/s to it. On the other hand, if Jupiter were at the position of the Earth, it would have produced a radial velocity as high as 28.4 m/s. Therefore, depending on the sensitivity of the present-day instruments, only giant planets, a few times more massive than Jupiter, orbiting very close to the star can be detected if the inclination angle is not too small. Till date, about 570 extra-solar planets in 428 planetary systems have been detected through the Radial Velocity method. Clearly, many of these planets are in planetary systems with more than one planet orbiting the parent star. In such case the periodicity of the shift in the absorption lines becomes complicated. Radial Velocity method can provide the orbital period of the planet. Since the orbital inclination of the planetary system is not known, Radial Velocity method can provide only the minimum possible mass of a planet assuming the system is viewed edge on or the inclination angle is $90°$.

The Transit Method

Many of you must have heard and some of you even witnessed a rare celestial event that took place on 6 June 2012. It was the last such event in this century. The event was a transit of the planet Venus across the Sun. It was visible as a dark circular spot on the solar surface which was moving slowly with time (Fig. 5.5). When the Moon passes through the straight line joining the Earth and the Sun, the sunlight is totally or partially blocked. We know this event as solar eclipse, total or partial, sometimes annular. When the sunlight is partially blocked by a planet, it is called a transit of the planet. We can observe the transit of only Mercury and Venus across the Sun because the Sun is the center of rotation of all the planets, and only those planets

Fig. 5.5 Transit of Venus (Credit: Author)

that are located inside the region between the Sun and the Earth can cause such an event visible to us in the Earth. If we were in Saturn, we could have observed the transits of Jupiter, Mars, Earth, Venus, and Mercury across the Sun.

However, a planet can transit another planet. If the "apparent size" of a planet that blocks the light of another planet is larger or smaller than the "apparent size" of the later planet, then the event is called occultation or transit, respectively. By "apparent size" we mean the size as seen by us and not the actual size. For example, since Venus is closer to us than Neptune, the apparent size of Venus is larger than the apparent size of Neptune, although in reality Neptune is much larger than Venus. So if Venus appears in between the line joining Neptune and the Earth, Neptune will be completely blocked by Venus. In that case we call it occultation of Neptune by Venus. Since the orbital distance of the planets and their sizes are fixed, both occultation and transit of a particular planet by another planet cannot occur. Venus always occults Neptune but it always transits Jupiter because the apparent size of Jupiter is larger than that of Venus. Transit or occultation of an object by a satellite is called eclipse. Therefore, when the Moon completely or partially blocks the Sun, we call it solar eclipse, total or partial, instead of calling it transit of Moon or occultation of Sun. Transit of a planet across its star is a very simple phenomenon—a tiny portion of the stellar disk gets blocked by the planet. So we see the shadow of the planet moving across the stellar disk. If during a transit we measure the brightness of the Sun or more precisely the total energy received from the Sun, then we find a slight reduction in the total energy. The amount of starlight reduced depends on the size and the distance of the planet. For a transit of Jupiter and Venus across the Sun, the decrease in sunlight would be just 1 %. But for the Earth, it would be as small as 0.008 %, i.e., about a thousand times less than that due to the transit of Jupiter. But this phenomenon plays a very important role in detecting extra-solar planets. When an extra-solar planet transits its parent star, the starlight diminishes albeit very slightly. Also, depending on the orbital period of the planet around its parent star, this phenomenon occurs periodically. There is no reduction in the brightness when the planet is behind the star. This method of detecting extra-solar planets by measuring periodic decrease and increase in the light of the star is known as Transit method. About 1,100 extra-solar planets in 350 planetary systems have so far been discovered by using the Transit method. This method provides more information about the planets than the Radial Velocity method can do. In the Transit method, astronomers measure the change in the brightness of the star by using an image photometer or simply a photometer for a few days in regular intervals. If the brightness of the star behaves like the one illustrated graphically in Fig. 5.6 with accurate periodicity, then that indicates the presence of a planet orbiting the star. The transit of the planet starts when the shadow of the planetary disk touches the stellar disk at position B. As the shadow enters the stellar disk, there is a sudden drop in the brightness of the star (point B to C).

Point B is the first contact for the shadow of the planet to the stellar disk and is known as ingress. The brightness of the star remains the same—in reduced amount, throughout the passage of the shadow across the stellar disk (C to E). Once the shadow of the edge of the planetary disk touches point E and crosses it, the starlight

Fig. 5.6 Illustration of the transit method (Illustration by the author)

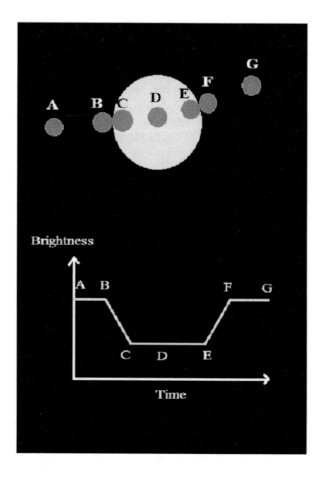

starts increasing, and after the last contact at point F, the brightness of the star becomes the same to the amount before the transit. The last contact point at F is called egress. The same process repeats again and again in regular intervals depending on the time taken by the planet to move around the star. When the planet is behind the star, it is completely blocked by the star and such an event is called secondary eclipse.

Although the Transit method is comparatively easier to employ and needs a simple instrument, it requires the planet to orbit the star almost edge on. Therefore, only those planets whose inclination angle is very high can be detected through the Transit method. Also, since a transit can occur due to another closed companion star in a binary star system, it usually needs the Radial Velocity method to confirm the presence of the planet. Nevertheless, Transit method can provide the size of the planet if the size of the star is known. While the Radial Velocity method can provide an estimation of the minimum mass of the planet, the Transit method can provide the exact mass of the planet through Kepler's third law if the mass of the star is known. This is possible because the inclination angle of the planet can be

Detection Methods

derived from the transit duration and the ingress as well as the egress time. Hence, Transit method can provide the mean density of the planet which is crucial in determining whether the planet is rocky or gaseous.

The Gravitational Lens Method

Till date, most of the extra-solar planets are discovered either by using the Radial Velocity method or by using the Transit method. But a few of them are discovered by using other methods as well. Gravitational micro-lens is one of them. The principle behind the gravitational lens method is based on Albert Einstein's General Theory of Relativity. According to this theory even light alters its path due to the gravitational attraction of a nearby massive object. Light from a distant star travels in a straight line and reaches the observer. However, if there is a massive object such as a star or a planet nearer to the observer and to the line joining the path of the light from the distant star to the observer, then the ray of the light from the distant star slightly bends towards the intervening star as shown in Fig. 5.7. This gives rise to gravitational lens effect which brightens the distant star. It is similar to the common experience of magnifying an object by using a convex magnification glass or lens. This effect is very significant when a galaxy lenses another galaxy or a massive black hole bends the light of a distant galaxy or quasar. Gravitational lens by small objects such as stars or planets is called micro-lens. This method also needs an image photometer to record the change in the brightness of a distant star continuously for a long duration. It takes about 2–3 weeks to record the complete event. But one cannot continue the observation during the day time. Therefore, a network of telescopes situated at different places around the world is used to record the brightness change. As illustrated in Fig. 5.7, when the distant star or the source of light moves towards the line joining the Earth and the lens object, the brightness of the star gradually magnifies with time. The brightness reaches a maximum value when the source crosses the straight line joining the lens and the observer at the Earth, i.e., when the three objects—the distant star, the lens star, and the Earth—are aligned in a straight line. Subsequently the brightness starts decreasing as the source

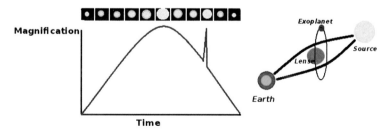

Fig. 5.7 Illustration of gravitational micro-lens of a distant star by an extra-solar planet around a nearby star (Illustration by the author)

moves away from the line joining the lens and the Earth. Now, if there is a planet orbiting the lens star, it also acts as a gravitational lens to the source. So, while the source is moving away from the line of sight and hence its brightness is also decreasing systematically, it again brightens up when it crosses the line joining the Earth and the planet. The second magnification occurs due to the planet acting as the lens. By recording such an event wherein a double gravitational lensing—a sharp but shorter one within the duration of the main event—occurs, astronomers confirm the presence of an extra-solar planet orbiting the lens star. However, this is a chance effect. The event occurs only once unless another distant star in the future crosses the line joining the Earth and the planet hosting star. But such a repeat is very rare. Till date a total of 29 extra-solar planets in 27 planetary systems have been detected through gravitational micro-lens method. Gravitational lens method can provide the mass of the planet accurately even if it is very small and located far away from its parent star. Also, the event can occur for any inclination angle of the planetary orbit. However, it cannot determine the orbital period or the size of the planet.

Astrometry

The presence of a massive planet around a star introduces a wobble in the star as discussed before. In the Doppler method, this wobbling or radial velocity of the star is detected and measured through the shift in the absorption lines that appear in the spectrum of a star. However, the wobble or the slight disturbance in the orbit of the star can be detected directly by imaging the star at different positions. This is known as astrometry. But, due to the effect of the Earth's atmosphere, measuring a small change in the position of the star due to the planet is extremely difficult by any ground-based telescope. Therefore, this method is used mainly by space-bound telescopes such as Hubble Space Telescope which has discovered a few giant extra-solar planets orbiting around very low mass stars such as M dwarf stars or Brown Dwarfs through astrometry. A heavy planet can give rise to significant radial velocity to a light star such as an M dwarf. On the other hand, in a binary star system, the orbits of the stars deviate from that estimated by Kepler's laws. This deviation indicates the presence of a third body in the system. Astrometry is the oldest method used by the astronomers in order to detect a small component of a binary stellar system. This method was used in 1855 by Captain W. S. Jacob of Madras Observatory to predict the presence of a planetary mass object in the binary star system 70 Ophiuchi.

Pulsar Timing

A pulsar is a rotating neutron star which is the end product of a very massive star, a few times more massive than the Sun. Any star heavier than about one and a half times the Sun ends into a neutron star. A pulsar is a rapidly rotating neutron star. Pulsars are powered by the conversion of rotational energy through the interaction of its intense magnetic field. Pulsars emit radio wave in a narrow angle through their magnetic poles. Since a pulsar rotates with an extremely accurate period, the radio wave from it arrives at the Earth in a regular interval of time just like the light observed from a light house near any sea or ocean. Now if a planet orbits around the pulsar, the interval at which the radio signals or the pulses arrive at the Earth gets modulated and the planet is detected from the anomaly in the pulse interval. Aleksander Wolszczan and Dale Frail used this method to discover the planets around the pulsar PSR 1257+12 in 1992. Planets around only a few pulsars are discovered till now. It is still not known how the planets around a pulsar are born.

Polarimetry

Light travels as a wave which oscillates in the direction perpendicular to the direction of propagation. Usually the light wave from a star oscillates randomly. But when this randomly oscillating light or "unpolarized" natural light from the star reflects from a planetary surface or gets scattered by the atoms, molecules, or dust particles present in its atmosphere, it starts oscillating in a particular direction. This phenomenon is called polarization of light, and hence, the reflected light of a planet is polarized. The polarization of light can be detected by using an instrument called polarimeter. The intensity of the "unpolarized" or natural light is the same in all directions, but the intensity of polarized light is different at different directions. If we see a ray or beam of polarized light, such as the light from a television or a mobile phone or the reflected light from a pond, through a polarizer, we notice the change in the brightness of the light when the polarizer is rotated. Many modern-day sunglasses act as a polarizer. Similarly, a polarimeter fitted with a camera onboard a telescope records the change in the brightness of a planet or a satellite at different directions and provide the amount of anisotropy or the degree of polarization of light. The amount of polarization of the reflected light would vary systematically depending on the planetary phase angle. The planetary phase angle is the position of the planet around the star with respect to the observer and varies from 0° to 360°. The portion of the planetary surface illuminated by the starlight varies at different phase angles. This is similar to the different phases of the Moon we observe on different days. Because of the change in the illuminated area of the planetary surface, polarization also changes during one orbital motion of the planet around the star. Therefore, the polarization is zero when the night side of the planet is faced towards the observer. However, when the day side of the planet is faced towards the observer, the planet is blocked by the star and hence no polarization can be detected. The amount of polarization observed should be the maximum when half of the planetary surface is illuminated by the starlight. Therefore, the amount of polarization of the reflected light should increase from zero to a maximum value and then reduces to zero during one orbital period. This will repeat according to the orbital period of the planet around the star. The light from a normal star is however unpolarized. Therefore, detection of polarized light of a star in a periodic pattern could indicate the presence of planets around it. However, since, the polarized reflected light of a planet is mixed up with the unpolarized or natural light of the star, the amount of polarization is extremely small. But if the planet is relatively distant from the star such that the starlight can be blocked and only the reflected light of the planet is received, then the amount of polarization should be significant.

This method can help in detecting planets that are orbiting even at low inclination angle. Polarimetry is the newest method proposed and yet to be applied in discovering any planet. This is mainly due to the lack of suitable instrument. In the near future, the Gemini South telescope at Chile will use a polarimeter and image the polarized light of extra-solar planets and even proto-planetary disks around stars by blocking the intense unpolarized starlight. Polarization is sensitive to the atmospheric composition of a planet, and hence, this method should be useful in understanding the chemical composition of the atmospheres of extra-solar planets. It is worth mentioning in this context that the presence of sulfuric acid in the atmosphere of Venus was discovered through polarimetry. Almost all the solar planets and a few satellites around them show polarization. The light of the Moon is also polarized. In future, large ground-based telescopes, such as the Thirty Meter Telescope (TMT) and the European Extremely Large Telescope (EELT), will use polarimetry for the study of the atmosphere of extra-solar planets. Since the atmosphere of Brown Dwarfs very much resembles the atmosphere of planets, it was predicted that the light from the warmer Brown Dwarfs or the L dwarfs would be polarized due to the scattering by dust cloud in the visible atmosphere. Subsequently it is confirmed that most of the L dwarfs are polarized. The coolest Brown Dwarfs, Y-dwarfs, should also be polarized due to light scattering by water cloud.

Other Methods

There are a few other methods such as imaging, Doppler isolation, nulling interferometry, etc. that are used to detect extra-solar planets. However, these methods are complicated and also difficult, and so only a few extra-solar planets are discovered by using such methods. Recently a few very hot and young extra-solar planets are directly imaged by blocking the light of the parent star. An instrument called "coronagraph" is used to block the starlight. In the future a large number of such planets are expected to be imaged directly.

After the specially designed space telescopes "Kepler" launched by NASA and "Corot" launched by European Space Agency (ESA), several extra-solar planets with different physical properties hitherto unknown to us were discovered through the Transit method. About 1,800 different kinds of worlds beyond our own have so far been discovered by various groups of astronomers. On the other hand, the recently launched space telescope GAIA is expected to discover thousands of extra-solar planets using the Transit method and astrometry. The progress in discovering extra-solar planets is overwhelming, and the variety of extra-solar planets discovered so far is beyond our imagination. We shall discuss about the nature of these fascinating planets in the next chapter.

Chapter 6
An Amazing Zoo of Planets

> *The diversity of the phenomenon of nature is so great, and the treasures hidden in the heavens so rich, precisely in order that the human mind shall never be lacking in fresh nourishment.*
>
> —Johannes Kepler
> (*In* Terra inest virtus, quae Lunam del)

How Much Do We Know About Them?

Prior to the space era, our knowledge regarding the properties of the solar planets was very limited. We knew about their orbital periods, sizes, masses, reflectivity, etc. and inferred their surface temperatures as well as the climates. But as soon as various space missions were launched and spacecraft such as Mariner, Viking, Voyager, Pioneer, etc. passed very near to the planets or orbited them closely and sent spectacular images of the surface of the planets, our knowledge increased enormously. Voyager I and Voyager II started their journey in the year 1977, and they have most probably crossed the boundary of the solar system a few months ago. It's not possible to send a spacecraft to another star or to another planetary system because of the vast distance. Fortunately, because of the advancement in technology, our present knowledge on the extra-solar planets is much better than it was on the solar planets prior to the space era. A combination of Radial Velocity method and Transit method provides an accurate measure of the mass of the planets. Transit method can measure the size of the planets, and hence, we can calculate the mean density of a large number of planets outside the solar system. The mean density provides an insight on the internal structure of the planet. Often astronomers take a spectrum of the star when the planet is transiting it, and then they take another spectrum during the secondary eclipse, i.e., when the planet is behind the star. A careful comparison of these two spectra provides the chemical composition, e.g., the various kinds of gases present in the atmosphere of the planet. During the secondary eclipse, i.e., when the planet is completely blocked by its parent star, a drop in the infrared light from the star is observed. The amount of infrared radiation decreased is the amount emitted by the planet. Therefore, the amount of radiation reemitted by the planet can be estimated from the secondary eclipse which in turn provides the thermal properties of the surface or the atmosphere. From the thermal properties, much about the atmospheric composition and structure can be inferred as the radiation interacts with the matter in the atmosphere before emerging out. On the other hand, a few planets which are young and hot and orbiting far away from their stars can be imaged directly by blocking the starlight. Since the planet is far away, at a distance similar to that of Pluto from the Sun, its light can be resolved from the starlight. The spectra of these planets are obtained directly. The spectra confirm the presence of water vapor, carbon dioxide, and methane molecules in the planetary atmosphere. Note that the light from this type of self-luminous planets is not the reflected starlight but their own light. The surface of a planet that reflects starlight is periodically illuminated partially or fully depending on its position or phase with respect to the star and the observer. But self-luminous planets always emit light from the whole surface. In that sense they are very similar to the Brown Dwarfs.

Size Matters: Rocky Versus Gaseous

Mass and size play an important role in determining the environment of the planets. Planets with large size such as Jupiter, Saturn, Uranus, and Neptune are gaseous. Their density is therefore low. The maximum mass of a planetary object is decided from the hydrogen-burning limit. For a solar composition, therefore, no planet can have mass more than 13 times the mass of Jupiter. If an object orbiting around a solar-type star is found to have mass more than 13 times the mass of Jupiter, it should be a Brown Dwarf component in a binary star system and not a planet. However, the maximum size of a planet is not well defined. The largest planet discovered till date is about double the size of Jupiter. Usually, the maximum size of a planet around any solar-type star should not differ much from the size of Jupiter. In other words, Jupiter is the limiting size of any planet around any solar-type star. It is usually believed that all planets that have a size similar to or larger than Neptune are gaseous. They do not have a solid surface or a solid core. If a lighter planet with smaller size is made entirely of gas, it may not be gravitationally stable. That is why we do not see Earth-size gaseous planet. The presence of rock or rocky surface can be inferred from the mean density of the planet. Usually planets with smaller size such as the Earth, Mars, Venus, and Mercury have rocky surface. Since they are close to the Sun, much of the gas was evaporated during their birth. On the other hand, the Jovian or the outer planets are far away from the Sun, and they have retained their gaseous components. As a consequence, the Jovian planets are much larger and heavier than the inner planets. However, in our solar system there is no planet intermediate to the Earth and Neptune. So we don't know any solar planet which is in the transition stage of gaseous and rocky planets. Planets slightly larger than the Earth could also have rocky interior, but their internal structure might be different from that of the Earth. These are called super-Earths. Planets smaller than Neptune and with mass not exceeding about ten times the mass of the Earth should have solid core and extended gaseous atmosphere. These are known as mini-Neptunes. Intermediate to mini-Neptune and super-Earth, there can be another type of planets—ocean planets with solid water layer covered by liquid water surface. Ocean planets should have solid mantle and core. The largest known rocky planet discovered till date is five times heavier than the Earth and about one and a half of the size of the Earth. On the other hand, there is no lower limit for the size of planets. Mercury, Moon, Pluto, Ceres, etc. are all rocky objects. The smallest extra-solar planet discovered so far is as small as the Moon. However, we do not know if we should consider these small objects as dwarf planets because the definitions of planets and dwarf planets given by IAU are confined to the solar system only. Comets and asteroids are also made of rocks.

Hot, Warm, and Cold

In addition to the size, the surface temperature too plays a crucial role in determining the atmosphere properties of a planet. The surface temperature of a planet can be derived from its orbital distance and the surface temperature of the star or the heat radiated by the star. The heat emitted from the surface of a star depends on the age and the mass of the star. Our Sun is a middle-aged yellow dwarf star and belongs to the spectral type "G." So the Sun is hotter than only K-, M-type stars and the Brown Dwarfs. Figure 6.1 shows the surface temperature of planets at different distances from the Sun or a Sun-like star. Although Mercury is nearer to the Sun, it has no atmosphere to absorb and retain the solar heat. On the other hand, Venus has a thick atmosphere which absorbs a good amount of the heat received from the Sun. Because of this, Venus which orbits the Sun at a distance of about 0.72 AU is hotter than Mercury although Mercury is closer to the Sun, at a distance of 0.4 AU. The surface temperature of a planet depends on the reflecting capacity of the atmosphere or the surface. The ratio between the energy reflected by the planet and the energy received at the surface of the planet from the star is known as Bond albedo or simply albedo. Therefore, albedo gives the fraction of starlight reflected by the planetary surface. The higher the albedo of a planet, the more light reflected from its surface and hence the less the surface temperature. A measurement of the albedo provides significant insight on the composition of the atmosphere. Although the surface temperature of a planet (known as equilibrium temperature) is determined from the brightness or luminosity or the surface temperature of the star, its orbital distance, and the albedo, the actual temperature inside the atmosphere of the planet is determined by the amount of reemitted infrared radiation that is trapped by the atmosphere. This is called the Greenhouse effect, and we shall discuss it in detail in Chap. 8. The equilibrium temperature of the Earth can be determined by using the

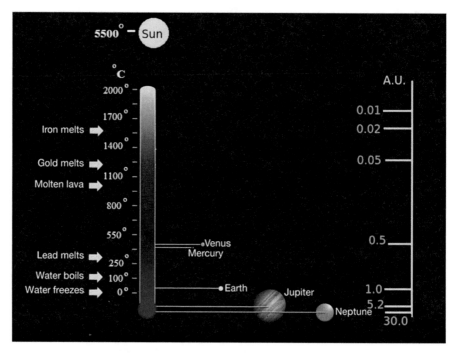

Fig. 6.1 Surface temperature scale of the planets with respect to their distance from the Sun. The temperature is given in Celsius and the distance is given in Astronomical Unit (AU) which is the distance between the Sun and the Earth. The surface temperature of the Sun is about 5,500 °C (Illustration by the author)

surface temperature of the Sun (5,505 °C), the mean distance of the Earth from the Sun (1 AU), and the albedo of the Earth (0.3), and it is −18 °C. But the actual mean temperature of the Earth is 15 °C.

The hottest planet in the solar system is Venus whose surface temperature is slightly less than 500 °C (about 900 °F) which is sufficient to melt lead. But outside the solar system, several planets are discovered that are orbiting very near to their stars and hence much hotter than Venus. Some of these planets have surface temperature exceeding even 3,000 °C. The hottest planet discovered so far has a surface temperature of about 7,000 °C. This planet, known as Kepler-70b, is orbiting a hot star of spectral type B in just 5.8 h. The surface temperature of the star Kepler-70 is about 27,500 °C, and the distance of the planet from such a hot star is just 0.006 AU which makes this planet super hot. A rocky planet very near to its parent star should not have an atmosphere because the intense heat from the star would cause complete evaporation of the gas. On the other hand, there is evidence that indicates a giant gaseous planet very near to its parent star, losing its mass due to rapid evaporation.

A Large Variety of Extra-solar Planets

Mercury, the planet closest to the Sun, is 0.4 AU away from the Sun. But several planets have been discovered which are orbiting as close as 0.01 AU from their parent star. Therefore, their surface temperature is several times higher than that of Mercury or Venus. On the other hand, several planets are discovered which are even ten times heavier than Jupiter and as large as Jupiter. Based on the size and the surface temperature of all these planets, we can divide them into several groups. Most of these extra-solar planets are quite different from the solar system planets. We know only five kinds of planets in the solar system. Mercury, many of the dwarf planets, and most of the large satellites are purely rocky and very small. They don't have any atmosphere. Venus, Earth, and Mars have solid surface and atmosphere. Only the Earth is a habitable planet. Jupiter, Saturn, and Uranus are icy cool giant gaseous planets. Neptune is, although gaseous, much smaller than them and only a few times more massive than the Earth. So, according to their sizes, the solar planets can be divided into four groups: subterrestrial (Mercury and Mars), terrestrial (Venus and the Earth), Neptunian (Neptune and Uranus), and Jovian (Jupiter and Saturn). Among the terrestrial planets, only the Earth is habitable, i.e., the surface temperature of the Earth is suitable for water to exist in liquid form. Mars is considered to be marginally habitable. Because of its highly elliptical orbit around the Sun, Mars becomes warm enough to melt ice when it comes closest to the Sun.

It is interesting to learn that till now a large number of planetary systems consisting of two to six planets orbiting the same star are discovered. Planets with different surface temperatures and of different sizes around the same star are discovered by using both Radial Velocity and Transit methods. However, not a single planetary system discovered so far resembles our own solar system. Nevertheless, by now so many different kinds of extra-solar planets are discovered that we need as many as 18 groups to describe them broadly on the basis of their size or mass and surface temperature. These 18 groups are made out of six different sizes or masses and three different ranges of surface temperature depending on their orbital distances and the brightness of the parent star. The six different sizes of planets are (1) sub-Earth or subterrain (smaller than the Earth), (2) Earth or terrain, (3) super-Earth or super-terrain, (4) mini-Neptune, (5) Neptunian, and (6) Jovian. The three temperature ranges are (1) hot, (2) warm including "habitable," and (3) cold. The cool planets should be far away from their stars, and therefore, it is hard to detect them. A few cool planets are discovered by using the gravitational micro-lens method. But this method cannot determine the size or the orbital period of the planet. However, there is another kind of planets that are far away from their stars, but they are very young and hot and so they have high temperature of their own. Let us now discuss about each of these different types of planets briefly.

The Hot Jupiters or the Roasters

The first confirmed extra-solar planet 51 Pegasi b (the star is designated as 51 Pegasi) belongs to the group of hot Jupiters or hot giant gas planets or roasters. They are literally roasted because of their close proximity to their parent stars. These planets can be as massive as 10–13 times the mass of Jupiter. They are composed mainly of hydrogen and helium gas. However, because the core is under tremendous pressure, the gas in the core should be in liquid form. During the first few years since the discovery of 51 Pegasi b, only this type of planets was discovered because they are very near to their stars, they are giant in size, and they are several times more massive than Jupiter. So they exert maximum radial velocity to the stars. At the same time, such a huge planet reduces the brightness of the star significantly while passing in front of the star. Therefore, it is comparatively easier to detect them through Radial Velocity method and Transit method. Since they orbit as close as 0.01 AU, their surface temperature is as high as 1,500–2,500 °C. If they were less massive, then at this temperature the gas could have escaped from their surface, and ultimately they could have completely evaporated. Even at a distance of about 0.4 AU from the Sun, Mercury had lost its gaseous atmosphere because it could not keep the gas bound by its low gravitation. A hot Jupiter or a roaster is so closed to its parent star that it is often tidally locked with the star. That is why such a planet completes one rotation around its own axis during the same period it orbits around the star. In other words, their one day extends to their one year. This phenomenon is similar to the planet Mercury. Our Moon is also tidally locked with the Earth, because of which one lunar day is about a month which is equal to the time taken by the Moon to orbit the Earth. That is why we always see the same side of the Moon.

Further observations of these planets indicated high speed storm occurring in their atmosphere. They should also have dust clouds and hazes in the upper atmosphere. The transit spectra of these planets show signature of water (of course in gas form), carbon dioxide, and other molecules in their atmospheres.

The discovery of a large number of Jupiter-size planets orbiting very near to their parent stars is a surprise to the astronomers because it challenges our conventional knowledge on planet formation. A planetary system with one or more giant planets orbiting very near to the star implies a sharp contrast to the configuration of our own solar system. In our solar system, smaller planets were born nearer to the Sun, and the gas giants were born far away from it. So is there a completely different mechanism through which such close-in giant gas planets are born around other stars? We do not know. At present, it is believed that these gas giants were actually born far away from their parent stars but they migrated very near to the stars during a few million years after their birth. However, there is no observational evidence supporting such planetary migration.

Some well-studied and remarkable extra-solar planets under this category are HD 209458b, HD 189733b, Kepler-5b, Kepler-6b, Kepler-7b, and Kepler-8b. All these planets are about 16 times larger than the Earth, and their surface temperature ranges from 1,700 to 2,200 °C. The transit spectra of some of these planets show carbon dioxide, water vapor, and methane molecules. There is indication of a storm in the atmosphere of HD 189733b. On the other hand, observation reveals evaporation of the atmosphere of HD 209458b due to strong stellar irradiation.

Not all the close-in hot planets or roasters are gas giants. Planets similar to the Earth and Neptune in size and mass are also found very near to many types of stars—solar type, faint dwarfs, etc. A few of such planets are quite interesting. They are discovered by the space telescope Kepler operated by NASA. The stars having planets are designated as "Kepler" followed by a number indicating the discovery sequence by the Kepler space telescope. For example, Kepler-22 means the 22nd star around which planets are detected by Kepler. The planets are designated with alphabets. For example, Kepler-22b is the first planet detected by Kepler around the star Kepler-22, Kepler-22c is the second one, and so on. Note that the alphabets do not imply the distance to the planet from the star as a series. So the planet designated by "b" could be farther from the star as compared to the distance of the planet designated by "c," but the planet "b" was discovered before the discovery of planet "c." Usually the planets closest to the star are discovered first, but there are exceptions to this.

Kepler is a spacecraft launched in March 2009. It has a telescope about half the size of the Hubble telescope. An instrument called photometer is used by this telescope to detect extra-solar planets through the transit method. It is directed to the constellations of Cygnus and Lyrae. Kepler can detect planets that are 30–600 times less massive than Jupiter. However, in order to detect a planet which is in the Habitable Zone of a solar type of star, it needs 3 years of observation because such planets orbit the star in a year, and it needs at least three transit events to confirm the presence of the planet. Besides Kepler, another space telescope, CoRoT (COnvection ROtation and planetary Transit), operated by the European Space Agency (ESA) also discovered a large number of extra-solar planets. Many planets, including 51 Pegasi b, are however discovered by the ground-based telescopes. Prior to the launch of the space telescope Kepler, we could detect planets around only those stars that are about 300 light years away from us. But Kepler can detect planets around a star as far as 3,000 light years from the Earth. Note that 1 light year is equal to about 9.5 trillion km. As of December 2013, Kepler has completed its successful mission. But the huge amount of data collected by Kepler is still under analysis. The stars suspected to have planets are designated as Kepler Object of Interest or KOI followed by a serial number. As soon as the presence of a planet is confirmed, the designation is elevated to Kepler followed by the serial number of discovery by Kepler telescope and an alphabet as described above. On the 18 December 2013, the European Space Agency has launched a telescope called GAIA which is supposed to take high-resolution images of a billion stars in our galaxy. GAIA should extend the discovery of extra-solar planets in great extent.

Hot Neptunes and Mini-Neptunes

While Jovian planets are 6–22 times larger than the Earth, the Neptunians are two to six times larger than the Earth in size. Planets under this category are also gaseous in nature. Their mass lies between the mass of Neptune and Uranus. Kepler-4b, HD 160691c, Gliese 436b, etc. are some of the remarkable extra-solar planets that belong to this category. Gliese 436b is about four times larger than the Earth but about 22 times heavier than it. It is suspected that the core of this planet may have solid rock. If so, then it should better be considered as a mini-Neptune. Since the parent star Gliese 436 is an M-type star, the coolest among all types of stars, the surface temperature of the planet, Gliese 436b is about 440 °C although it is orbiting the star at a distance as close as 0.03 AU. The temperature of Gliese 436, a red dwarf star, is about 3,300 °C, almost half of the surface temperature of the Sun. Although the atmosphere of Neptune in the solar family contains ice, the atmosphere of hot Neptunes such as Kepler-4b and Gliese 436b is entirely different. Kepler-4 is a solar type of star with a surface temperature very similar to that of the Sun. But it is slightly larger in size and heavier than the Sun. The planet Kepler-4b is orbiting it at a distance of about 0.046 AU. Therefore, the surface temperature of this planet is as high as 2,000 °C. This planet is about 25 times heavier than the Earth and about four times larger than the Earth. Therefore, this planet may also belong to the group of hot Neptunes. Mini-Neptunes, hot or cool, are not present in our solar system.

In 2011, Kepler discovered as many as six extra-solar planets around the solar-type star Kepler-11. All the planets around Kepler-11 orbit within a radius of 0.5 AU, and so all of them are very hot. Except Kepler-11b, all the planets around Kepler-11 are Neptunians. Kepler-11b is slightly less than double the size of the Earth, but the radii of Kepler-11c, d, e, f, and g range from 3 to 4.5 times the radius of the Earth. The masses of Kepler-11b, c, d, e, and f are estimated to lie between twice the mass of the Earth and the mass of Neptune, but their densities are so low that except Kepler-11b, none of them could have a solid surface or rocky core and therefore may be considered as hot Neptunes. Kepler-11b may however belong to another class of planets, the super-Earths, and we shall discuss about this type of planets in the next section.

Sub-Earths, Earths, and Super-Earths

By sub-Earth, Earth, and super-Earth, we mean extra-solar planets whose sizes are less than, similar to, or slightly greater than the size of our Earth, respectively. The size of a super-Earth varies from 1.25 to 2 times the size of the Earth. Also their mass is comparable to the mass of the Earth. The super-Earths are two to six times heavier than the Earth, while sub-Earths have mass less than that of the Earth. Note that by an Earth-size planet, we mean a planet with size ranging between 0.75 and

Sub-Earths, Earths, and Super-Earths

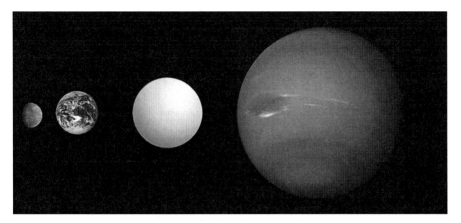

Fig. 6.2 Sub-Earth, Earth, super-Earth, and Neptune: a comparison of size (Credit: Reconstructed from NASA images)

1.25 times the size of the Earth. All these planets are rocky and their internal geology is similar to that of the Earth. However, their climates may drastically differ from that of the Earth, and some of them may not even have an atmosphere. Therefore, not all of them are Earth-like planets. These planets are also called subterrain, terrain, and super-terrain or subterrestrial, terrestrial, and super-terrestrial (Fig. 6.2).

The interior of the Earth or a terrestrial planet can be divided into three distinct regions (see Fig. 6.3). (1) The crust which is on average 30 km thick. It consists of basalt rocks under the oceans and of silicate granite under the continents. (2) The mantle which is about 2,860 km thick and consists mostly of a mineral called Perovskite. Perovskite is a dark and dense rock composed mainly of iron and magnesium. This mineral is about 50 % denser than granite. (3) The core which is made of pure iron and alloys of iron. The temperature of the core ranges from 5,000 to 7,000 °C under high pressure, and so the whole core is in liquid form. Just at the bottom of the mantle and just above the core, there is a layer of material, known as post-Perovskite. This layer is about 150 km thick for an Earth-size planet. Post-Perovskite is a crystalline form of Perovskite under tremendous pressure.

A super-Earth should also have a crust, mantle, and core. But the geology of a rocky super-Earth differs from that of the Earth because the pressure inside them is so high that the entire mantle consists of post-Perovskite instead of Perovskite. Now if a super-Earth contains liquid water or ocean at the surface, the crust is replaced by a thin layer of liquid water and a thick layer of solid water beneath it. Although the temperature of this region is too high to keep water in liquid state, the high pressure makes water to exist in solid form. This high pressure water ice which can exist at high temperature is known as ice vii, ice x, and ice xi. This type of extra-solar planet is called water super-Earth or ocean super-Earth.

On the other hand, the sub-Earths or the subterrains are smaller than the Earth and less massive. Their interior is similar to the Earth's interior, but as the density

Fig. 6.3 Interior of rocky Earth, super-Earth, and ocean or water super-Earth (Illustration by the author)

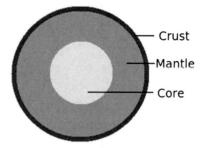

Rocky Earth and Super-Earth

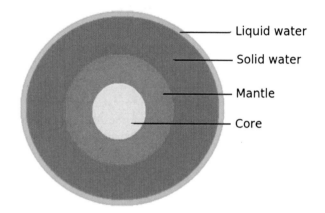

Water Super-Earth

reduces, the Perovskite layer in the mantle becomes thinner and Olivine becomes the main constituent of the mantle. Although not much is known about the interior of the Moon, it is expected that the entire mantle of the Moon or Pluto is made of Olivine and not Perovskite. Therefore, the interiors of sub-Earths, Earths, and super-Earths are characterized by the phase transitions from Olivine to Perovskite and to post-Perovskite. In fact our Earth falls between terrestrial planets or terrains and super-Earths or super-terrains because it has a thin layer of post-Perovskite. Astronomers have detected all these three types of planets outside the solar system. Although the solar system has the terrestrial and the subterrestrial planet, it does not have any super-terrestrial planet or super-Earth.

Since super-Earths are relatively large in size, their detection is comparatively easier. So the number of super-Earths discovered till date is greater than the number of terrestrial or subterrestrial planets discovered. One of the notable super-Earths is Gliese 581e which is twice as heavy as the Earth. It was discovered in 2009. Gliese 581 planetary system has another super-Earth, Gliese 581g which is three times heavier than the Earth. The planet COROT-7b (discovered by the space telescope CoRoT) is about five times heavier than the Earth and was also discovered in 2009. In Table 6.1, we provide a list of notable subterrestrial, terrestrial, and super-

Table 6.1 A few super-Earths, Earths, and sub-Earths detected by Kepler

Designation	Mass in Earth mass	Radius in Earth radius	Orbital distance in AU	Temperature (°C)
Kepler-10b	4.56	1.416	0.0168	1,550
Kepler-11b	4.3	1.97	0.09	630
Kepler-18b	6.9	2.0	0.0447	900
Kepler-36b	4.45	1.49	0.115	700
Kepler-37d	Not known	1.99	0.21	190
Kepler-42b	2.73	0.78	0.0116	250
Kepler-65b	Not known	1.42	0.035	1,300
Kepler-65d	Not known	1.52	0.084	750
Kepler-68c	4.8	0.953	0.09	Not known
Kepler-20e	0.39–1.67	0.87	0.05	760
Kepler-20f	0.66–3.04	1.034	0.11	425
Kepler-37c	Not known	0.742	0.1368	300
Kepler-42c	2.06	0.73	0.006	450
Kepler-37b	0.01	0.303	0.1	425
Kepler-42d	0.9	0.57	0.0154	175

Planets in the first nine rows are super-Earths, planets in the next four rows are Earths, and planets in the last two rows are sub-Earths. Unknown mass implies that only the maximum mass of the planet is estimated

terrestrial extra-solar planets discovered by Kepler. The planet Kepler-37b is slightly larger than our Moon but smaller than even Mercury. It is about one third the size of the Earth. Therefore, this planet may be considered as dwarf planet such as Pluto or Eris. Kepler-37c is slightly smaller than Venus, but Kepler-37d is double the size of the Earth. Therefore, this planetary system around Kepler-37 has all the three types of planets—sub-Earth, Earth, and super-Earth. However, the masses of Kepler-37c and Kepler-37d are yet to be determined. The star is similar to our Sun.

Besides Kepler, other telescopes such as CoRoT, HARP, VLT, Gemini, etc. have detected a good number of super-Earths. The sub-Earths are also detected through gravitational micro-lens and pulsar timing methods. It is expected that in the near future a large number of Earth-size planets will be detected by the space telescopes. However, as we can see in Table 6.1, most of these planets are orbiting very close to their stars, and hence, they are very hot. Therefore, although they resemble the Earth in their structure, they are not at all habitable. Smaller planets very near to their stars may not even have an atmosphere because the gas of the atmosphere evaporates due to the strong irradiation by the star. The warm sub-Earths, Earths, and super-Earths should resemble our own planet, the Earth. However, the detection of such warm terrestrial planets around a solar-type star is difficult by the existing facilities, and so the search at present is confined around cooler and dimmer stars.

Planets in the Habitable Zone

The term "Habitable Zone" was coined by astrobiologist James Kasting well before the first extra-solar planet around a normal star was confirmed. As the distance from the star to the planet increases, the amount of heat energy received by the planet reduces, and hence, the planet becomes cooler as seen in Fig. 6.1. In a planetary system, the region at which a planet should have a surface temperature higher than 0° but lower than 100 °C, i.e., a region where the surface temperature of a planet would allow water under normal pressure to remain in liquid state, is called the Habitable Zone or the Goldilocks Zone. Clearly, this depends on two factors: (1) the brightness of the star and (2) the distance to the planet from the star. However, the planetary albedo as determined by the chemical composition of the planetary atmosphere or the surface decides if the planet is cool enough to contain liquid water. The intensity of radiation of the star falls by the square of the distance. For the Sun or for a solar-type star whose surface temperature is about 5,500 °C, the inner edge of the Habitable Zone at which the surface temperature of the planet would be just less than 100 °C—the boiling point of water under normal pressure—lies at about 0.57 AU, slightly beyond the distance to Venus from the Sun. The outer edge of the Habitable Zone wherein the surface temperature of a planet should be just above the freezing point of water, i.e., 0 °C, extends up to the orbit of Mars. For a cooler star such as an M dwarf, the inner and outer edges of the Habitable Zone should be situated much closer to the star. As discussed earlier, the atmospheric composition of a planet plays a crucial role in determining the surface temperature of a planet. Water and carbon dioxide play important roles in maintaining the atmospheric temperature. Depending on the albedo which is a measure of the amount of the light reflected from the surface, the temperature of the atmosphere may vary significantly. Thus, a few planets which appear to be situated near the outer edge of the Habitable Zone may be actually cooler and inhabitable. However, this difference is not very wide and the capacity of the atmosphere to trap the reemitted infrared light may push them into the Habitable Zone. In that sense the Habitable Zone is a loosely defined term. Further, the brightness of a star changes with time. The Sun was about 30 % fainter during its birth. Therefore, the age of the star too plays an important role in determining the Habitable Zone.

It is not necessary that all the planets in the Habitable Zone are terrains or super-terrains. They could be Jupiter- or Neptune-like gaseous planets. Hence, even if they are in the Habitable Zone, they cannot be habitable because they do not have a rocky surface to retain liquid water. A few planets larger than a super-Earth and smaller than Neptune are discovered in the Habitable Zone of their stars. They can be considered as warm mini-Neptunes. Kepler-22b, discovered in December 2011, is one of such planets. It is about 2.4 times larger than the Earth, and hence, most likely, it does not have a rocky surface to retain liquid water, but it is found to be in the Habitable Zone of the star Kepler-22. Similarly, the extra-solar planet HD 85512b orbiting at a distance of 0.26 AU around a K-type star (temperature about 4,500 °C) should have a surface temperature of about 20 °C. But it is about four times heavier than the Earth. The extra-solar planet PH2b (Planet Hunter 2) discovered by a group of amateur astronomers is as big as Jupiter in size but located in the Habitable Zone. The planetary system around the solar-type star Kepler-90 has as many as seven planets. This system may possibly have many more planets which are yet to be detected. The seventh or the outermost planet Kepler-90h in this planetary system is in the Habitable Zone of the star. It orbits the star in 331.6 days. The surface temperature of Kepler-90 is about 6,000 °C, and the planet is 1.01 AU away from the star. Hence, the surface temperature of the planet should be about 20 °C. But the planet is as large as Jupiter and hence must be gaseous. It is speculated that a rocky moon of these planets, if any, may be habitable. But detecting such a satellite around a planet is difficult and beyond the scope of the present observational facilities.

A rocky planet in the Habitable Zone is considered to be a habitable planet. The most significant discovery of a rocky extra-solar planet in the Habitable Zone of a solar-type of star is reported during 2013 by the Kepler team. Kepler-69c is a planet orbiting at a distance of 0.762 AU around a G-type star Kepler-69 which is about 2,700 light year away from the Earth. The surface temperature of Kepler-69c should lie between 8 and 26 °C. It is 1.54 times larger than the Earth, and so it should be considered as a super-Earth with rocky surface. Besides Kepler-69c, Kepler-62, a K type of star, slightly dimmer than the Sun, has two planets Kepler-62e and Kepler-62f in its Habitable Zone. Kepler-62e is orbiting at a distance of 0.427 AU from the star, and it is 1.61 times larger than the Earth. On the other hand, Kepler-62f is 0.718 AU away, at the outer edge of the Habitable Zone of the star, and about 1.41 times larger than the Earth. Therefore, both Kepler-62e and Kepler-62f are super-Earths and should have rocky surface. At the same time they should have appropriate surface temperature for water to exist in liquid state. So far, these three habitable planets—Kepler-69c, Kepler-62e, and Kepler-62f—are the best candidates to search for life. Very recently, a few habitable planets around some M dwarf stars are discovered from the analysis of Kepler's data. However, the Habitable Zone of the faint M dwarf stars is so closed to the star that the planets within this zone may be exposed to intense ultraviolet and X-rays from the star. Therefore, it is doubtful if a rocky planet can be habitable in the Habitable Zone of M dwarfs. In other words, M-type stars may not have a Habitable Zone at all.

In the near future, several rocky planets in the Habitable Zone of their parent stars will certainly be discovered. However, the habitability of the planets, as discussed above, is determined by just two parameters—surface temperature and density. Although these two important properties provide necessary information that enables us to decide the habitability of a planet, they are not sufficient to confirm if all of these habitable planets can actually harbor life. The origin and evolution of life need a combination of many more favorable conditions. We'll discuss it in the next two chapters.

Directly Imaged Extra-solar Planets

All the planets that we have discussed in the previous sections—gas giant planets, Neptunian planets, super-Earths, etc.—are orbiting so closed to their parent stars that they can never be seen or imaged directly by any telescope, however powerful it is. This is because of the fact that the light from these planets cannot be separated from the intense starlight. Also, these planets are strongly or moderately illuminated or irradiated by the stars. They don't have much internal energy of their own, and they are heated by their parent stars.

In April 2004, the Very Large Telescope (VLT) at Paranal Observatory in Chile operated by the European Southern Observatory for the first time took the image of a planet outside the solar system. The image was taken in infrared light. The starlight was blocked by using an instrument called coronagraph. This has been possible because the planet is far away—at a distance of about 40.6 AU from the star which happens to be a Brown Dwarf designated as 2MASS 1207. The planet 2MASS 1207b (Fig. 6.4) is about four times heavier than Jupiter. This is an interesting system in the sense that the primary object could not become a star and the secondary one could not become a Brown Dwarf because of insufficient mass. Soon after the discovery of 2MASS 1207b, four similar hot and self-luminous

Fig. 6.4 The first directly imaged extra-solar planet (*red*) around a Brown Dwarf 2MASS 1207 (Credit: NASA and ESO)

planets were discovered around a star of spectral type A. The star HR 8799 is about five times brighter than the Sun. The star and the planets are quite young, only about 30 million years old, and the system is located at about 130 light years away from us. All the four planets are very similar to the planet 2MASS 1207b. The mass of the planets ranges from 6 to 9 times the mass of Jupiter. Since they are very young, they are still contracting under their gravitational pull, and hence, they are very hot by their own. Their surface temperature ranges from 600 to 800 °C. The planet closest to the star, HR 8799e, is about 14.5 AU away from the star, and hence, it takes about 45 Earth years to complete one orbit around the star. The furthest one, HR 8799b, is 68 AU away from the star and takes about 460 Earth years to complete one orbit. As a consequence, they hardly receive heat from the star. But they are heated by themselves. Detailed studies have revealed the presence of carbon monoxide, water, etc. in their atmosphere. They should also have silicate cloud in their atmosphere. In that way, the atmospheres of these planets are very similar to the atmospheres of Brown Dwarfs. One of them, HR 8799c, is found to be rotating very fast around its own axis. How a planetary system with so many giant planets was formed is still a mystery. Apart from 2MASS 1207b and HR 8799b, c, d, and e, a few more extra-solar planets have been directly imaged, Beta Pictoris b being one of them. Since these planets are formed only recently, a few tens of a million years ago, the dispersing proto-planetary disks around the stars are also observed. Therefore, they provide a good opportunity for the astronomers to understand how the planets and the planetary systems are born. In the near future, Gemini Planet Imager (GPI) at Gemini South telescope will discover and image several of such young but fascinating extra-solar planets.

Diamond Planets

Diamond is made of carbon. Under intense pressure and high temperature, the atoms of carbon form a specific lattice giving rise to a crystalline material which is known as diamond. So diamond is called as an "allotrope" of carbon. Diamond can be produced both naturally and artificially. The mantle of the Earth, at about 150–200 km beneath the surface, has the required pressure and temperature to convert carbon into diamond. It comes out to the surface of the Earth through volcanoes. Natural diamonds are quite costly because of their high light reflecting property and hardness. Slight impurity provides different colors to diamond. Just a 0.2 g (1 carat) of diamond costs about a few thousand US dollars. So what will be the cost of a planet slightly bigger than the Earth if it is made of diamond and that too is natural diamond? Can there be such a planet in the universe? The answer is yes, part of a whole planet could be made of diamond. The cloud out of which the Sun and the solar planets were formed had oxygen about double the amount of carbon. Therefore, in the Sun and in the solar planets, we find more oxides, such as water, carbon dioxide, silicate oxide, etc. The mantle of a rocky planet in the solar system is made of molecules formed by oxygen and silicon atoms. But if the initial star forming cloud consists of equal amount of oxygen and carbon, then the star would be carbon rich and any planet around it should be composed mainly of carbon. These are called carbon stars and carbon planets. Many such stars are known to us. 55 Cancri is such a carbon star about 40 light years away from us. But this star is special because a super-Earth, 55 Cancri e, is orbiting it at an orbital distance of 0.015 AU. The mass of this planet is seven to eight times the mass of the Earth, and it is about double the size of the Earth. Since it is orbiting very near to its star, it has a very hot surface with temperature above 2,000 °C. If the planet is rocky, then its interior should have ideal condition to convert carbon into diamond. In such case, the entire mantle of this planet should be made of diamond. This is known as "diamond planet." This planet was discovered in 2011. Such planets can be considered as the jewels of the sky. A diamond planet around a pulsar was discovered before the discovery of 55 Cancri e, and in future many more diamond planets should be detected making our galaxy quite costly.

The most interesting fact is that these diamond planets can have their owners. They can be habitable if positioned in the Habitable Zone of their parent carbon star. A "diamond planet" in the Habitable Zone may harbor life if it is supported suitably by other astronomical and geological conditions. Although such a carbon planet is not sufficiently heated by its star, the high pressure inside it may convert the carbon core into diamond.

Planets with More than One Sun

The diversity of planets is not restricted to the above-discussed extra-solar planets but is extended far beyond our imagination. We know that a large number of stars in our galaxy are binaries or belong to multiple stellar systems. In a binary system of stars, the heavier one is called the primary component, and the lighter one is called the secondary component. Circumbinary planets are orbiting such binary stars or stars in multiple stellar systems. Imagine a planet with two suns. Its day and night should be determined by the rise and set of two stars. Its climate should be determined by the radiation from two stars. Are there such planets in our galaxy? The answer is yes. Some of them are discovered by the Kepler telescope. Kepler-16A and Kepler-16B are two stars in such a binary system located about 196 light years away from the Earth. While the primary star Kepler-16A is a K-type star, the secondary is an M-type star. They orbit each other at a distance of only 0.2 AU and take 41 days to revolve around each other. Kepler-16b, a Saturn-size planet, orbits both the stars at a distance of about 0.7 AU. It takes about 229 Earth days to complete one orbit around the stars. Since the planet is far away from the cool stars, it should be in a frozen state. This planet is about three times lighter than Jupiter. Kepler-34b however orbits around a binary system consisting of two similar stars of spectral type G. The two Sun-like stars orbit each other in 27 days, while the planet orbits the system in 288 days. This planet is about five times lighter than Jupiter and slightly larger than Saturn in size. Although it is 1 AU away from two solar-type stars, the combined radiation of two equally bright stars makes the planet much hotter than the temperature of a planet in the Habitable Zone. The extra-solar planet Kepler-35b is very similar to Kepler-34b. It rotates around a binary system of two solar-type stars at a distance of about 0.6 AU and takes about 131 days to complete one orbit around the stellar system while the stars take 21 days to rotate around each other. On the other hand, Kepler-38b is orbiting a binary system consisting of a Sun-like star and a faint M-type star at a distance of 0.46 AU. The planet takes about 106 days to complete an orbit, while the stars rotate in 8 days around each other. Kepler-38b is a Neptune-size planet. Multi-planetary systems around binary stars are also discovered by the Kepler telescope. Kepler-47b, a gas giant of Jupiter size, and Kepler-47c, a Neptunian planet, are orbiting the binary stars Kepler-47A and Kepler-47B. It is possible that the system has more planets but not yet detected.

The above mentioned planets are orbiting two stars. Recently it is found that the closest star to the Sun, Alpha Centauri B, has a planet orbiting it at a distance of 0.04 AU. This planet is designated as Alpha Centauri Bb. Alpha Centauri is a triple star system consisting of two closed stars—Alpha Centauri A, a solar-type star as the primary, and Alpha Centauri B, a K-type star as the secondary. They are jointly known as Alpha Centauri AB. A third star Alpha Centauri C or Proxima Centauri is orbiting Alpha Centauri AB. The planet is orbiting around the secondary star Alpha Centauri B. Since this is the nearest star to the Sun, Alpha Centauri Bb is the nearest extra-solar planet to the solar system at about 4 light years away from us. This planet has not one or two but three parent stars that determine its day, night, and climate. Although it is extremely difficult to detect a planet around such a multiple system of stars, in future it may be possible to find out if there are any more planets orbiting this triple stellar system which is our closest neighbor.

Planets with No Sun: Rogue Planets

While a few planets are found to have more than one parent, other few planets are found to be floating freely, without having any parent. The origin of such planets is still an enigma. Either they are detached from their parent stars due to some reason or they are formed like a Brown Dwarf. However, they are not Brown Dwarfs because they lack sufficient mass to ignite even deuterium in their core. They are planetary mass objects. Should we call them sub-Brown Dwarfs? IAU has proposed so but it is debatable because in that case we need evidence to prove that they are formed by the core collapse of interstellar cloud—just like a Brown Dwarf or a star is formed. On the other hand, there is no way to know if they were formed through the proto-planetary disk around a star and then got detached or ejected out from a planetary system. Whatever be the mechanism of their formation, since they are planetary mass objects, we must at present consider them as planets—free-floating planets or rogue planets. PSO J318.5-22 is such a free-floating planet discovered in October 2013 by the Pan-STARRS 1 wide-field survey telescope and subsequently confirmed by other telescopes. This object is about 80 light years away from us. It is

Fig. 6.5 The free-floating or rogue planet CFBDSIR J2149-0403 (Credit: ESO/P. Delorme. Source: www.eso.org/public/images/eso1245c)

six times heavier than Jupiter and only 12 million years old. Because it is so young, it emits heat through the release of gravitational potential energy and so is visible. Thus, this planet is just like some of the directly imaged planets we have discussed before, but it is not orbiting any star. Another rogue planet was discovered during November 2012 by the French astronomers, and it is named as CFBDSIR J2149-0403 (Fig. 6.5). This planetary mass object is about 130 light years away from us and about 4–7 times heavier than Jupiter. Since the first discovery of such a free-floating planet, Cha 110913, it is expected that a large number of such planetary mass objects—planets or sub-Brown Dwarfs—are floating freely in the space. But they are so faint that it is extremely difficult to detect them. Astronomers at Pennsylvania State University discovered the planet Cha 110913 in the year 2004 by using the space telescope Spitzer, and subsequently the discovery was confirmed by Hubble Space Telescope. Such free-floating planets pose a challenge to the astronomers. If such a freely floating object approaches very near to a star and then gets captured by the gravitational attraction of the star, we may find a planetary system in which the planet is composed of matter with elemental abundance entirely different from that of its parent star. For example, if we discover a carbon or diamond planet around a solar-type star, then it has to be a free-floating planet either formed like a Brown Dwarf or was ejected from the planetary system around a carbon-rich star and then got attached with a solar-type star. Nature provides all types of possibilities to amaze us. But the most amazing planets will be the ones that are not only habitable but also harbor life and so will show clear signature of life.

Chapter 7
Life: A Delicate Process

> *O, wonder!*
> *How many goodly creatures are there here!*
> *How beauteous mankind is!*
> *O brave new world,*
> *That has such people in't!*
>
> —William Shakespeare
> *(In* Tempest)

In the previous chapter we have discussed that out of all the planets discovered or yet to be discovered, only those that are within the Habitable Zone of the star should have the possibility of harboring life. In the Habitable Zone, the planet should have appropriate temperature for liquid water to exist. But not all planets in the Habitable Zone can harbor life. The planet must have a solid surface to retain liquid water. So a rocky planet within the Habitable Zone of its parent star is considered as a habitable planet. Therefore, appropriate size and temperature are the two necessary conditions. But are these two conditions sufficient? We can immediately realize that it is not. The planet must have a sufficiently large atmosphere with enough free oxygen molecules for respiration, enough carbon dioxide for photosynthesis, etc. What about radiation environment, acidity, toxicity, etc.? Can life survive under intense ultraviolet radiation? Can life survive under frequent bombardment of comets and asteroids? Can life survive under an extreme seasonal change?

One can imagine that in other planets life could be made of such materials that it can withstand high temperature, can survive without oxygen and without organic nutrients, etc. Obviously, the biochemistry in that case will be completely different. But can we consider such a process in other planets as life? This takes us to a fundamental question—What is life?

Is There a Definition?

We know that the laws of physics are universal, i.e., they are same everywhere in the universe. Unfortunately, we do not know whether the laws of biology are universal or not. This is because of the fact that the only sample of life available to us is on the Earth. We are yet to find life beyond the Earth. We know only terrestrial life. Unlike the Big Bang event that created matter and energy at one instant and distributed everything in a homogeneous and isotropic manner everywhere in the universe, life is not originated everywhere in the universe at the same time or out of a single event. Our knowledge and understanding of biology is limited to the Earth. Therefore, our search is limited to life similar to the terrestrial life which is carbon based. A search for extraterrestrial life made of different materials and governed by different chemical processes will pose the question if we can term it as life. This situation arises because life is a system and there is no fundamental definition of life. Now what do I mean by a fundamental definition? If 400 years ago someone were asked what water was, one would have replied that water was colorless, odorless, tasteless, transparent liquid which flows from up to down, freezes at $0°$, and vaporizes at 100 °C, etc. This is a description of the physical properties of water. A description is always incomplete and so controversial. One can list a few substances such as alcohol, vinegar or mineral oils that have similar properties. However, if the same question is asked today, one would have answered that water is a molecule of two hydrogen atoms and one oxygen atom symbolically represented by H_2O. From this later statement, all properties of water can be determined univocally and without any ambiguity, and one can clearly distinguish water from any other colorless, odorless liquid. Unfortunately, a similar statement or definition for life is not possible because life is a system as a whole. Therefore, one has to resort to the description of various properties of life which can never be complete, consistent, or comprehensive. For example, NASA has adopted a "working definition"—Life is a self-sustaining chemical system capable of Darwinian evolution. Obviously, this definition or description cannot be accepted by all. For example, a biologist may argue that an embryo is very well a living entity but it's not self-sustaining. True, but an astronomer is not going to search or find out an embryo in other planets. If the astronomers can find the mother or the host of the embryo, it's sufficient to conclude that life exists. Similarly, just one or two cats or dogs may not be capable of Darwinian evolution. But for an astronomer, finding just one cat or dog in a whole planet will imply that either someone brought it from other place and the cat or the dog has survived there or there must be many more cats and dogs and other species to be found out in the same planet. Nevertheless, it is always better to compare a group of living entities with a nonliving entity and determine the minimum number of basic differences between them in order to recognize or differentiate life instead of attempting to find out a comprehensive description or a fundamental definition of life. A nonliving entity may be considered as a substance which can never change its physical and chemical properties in a short period without the influence of any external physical or chemical process. Once again, this is not a complete description or definition of a nonliving entity.

Terrestrial Life: A Complicated System of Chemical Processes

In the previous section we have mentioned that our knowledge on life or biology is confined to terrestrial life. Therefore, our attempt will be to find out similar terrestrial life in other planets. If life exists in any other form, we may not be able to distinguish it as life unless it is extraterrestrial intelligence that can identify itself as life. Therefore, the challenge to the mankind as a whole and to the astronomers in particular is to find out the existence of life in other planets which is similar to that in the Earth. In other words the immediate question is—Is anybody similar to us out there? Of course, if we detect rats, cats, dogs, and monkeys in another planet, we may not need to bother for anything. But that may not be the case. A planet can have life in many other different forms. We have already mentioned about the "working definition" of life that NASA has adopted. If we expand and elaborate it slightly, then we can describe life minimally in the following way.

Life is an ensemble of organic materials, governed by a self-sustaining system of chemical processes and characterized by (1) metabolism, (2) replication, and (3) evolution through natural selection. Organic materials are molecules made of carbon and hydrogen atoms. That is why terrestrial life is carbon-based life. Metabolism is a process by which a living entity converts matter (food) into energy through oxidization. Replication is a process by which a living entity reproduces and passes on information from one generation to other. Evolution by natural selection is response to the surrounding environment and adaptation.

One can add many more descriptions, but the above three are the basic characteristics of a living organism on the Earth. Many of the other properties such as growth, movement, death, etc. can be derived from the above three. Of course, one cannot claim that the above three characteristics are sufficient and can describe life adequately. Nevertheless, probably for the astronomers looking for life outside the Earth, this could be a sufficiently complete or working definition.

We know that the living species on the Earth have a wide range of size—from bacteria to blue whale. They all are made of "cell," the basic building block of life. A cell is the smallest structural unit which is functional. A group of similar cells form a tissue and several tissues join together to form an organ. Cells were discovered in 1665 by Robert Hooke. Cells are composed of billions of molecules, and they differ in size, shape, and function. While most of the bacteria are unicellular, plants and all developed animals are multicellular.

Terrestrial life however is not divided by their size, food habits, or the condition under which they live. Rather it is divided into two categories—prokaryotes and eukaryotes, depending on the nature of the cells that build them. All multicell terrestrial organisms are built by eukaryotic cells which are about ten times larger than the prokaryotic cells. Bacteria and Archaea are prokaryotes, while plants and all animals are eukaryotes. The multicellular life has emerged about 1.5 billion years ago, much after the unicellular life originated. But the Earth clearly is dominated by microbes. All cells contain genetic material in an encapsulated region which stores the information. For prokaryotic cells, the non-membrane encapsulated region that contains the genetic material is called nucleoid. In eukaryotic cell, the genetic material is contained in a membrane-bound capsule called nucleus.

The main organic molecules that play important role in an organism are (1) proteins, (2) carbohydrates, (3) lipids or fats, and (4) nucleic acids. We usually know proteins, carbohydrates, and lipids as the nutrients in our food. They are actually large and complex organic compounds made of carbon, hydrogen, oxygen, nitrogen, sulfur, etc. Protein plays an important role in the functioning of organisms. Many of the hormones that act as messengers consist of proteins. Enzymes that catalyze chemical reactions are made of proteins. Hemoglobin in blood that carries oxygen is also made of proteins. Proteins are made of amino acids. Carbohydrates such as glucose and sucrose store the energy and play important role in metabolism. Lipids or fats form the membrane of the cells. One of the nucleic acids—deoxyribonucleic acid, popularly known as DNA—forms the gene that carries the information from one generation to another generation and responsible for replication through cell division. Each specimen of organism shares a set of common information within the DNA. DNA has a double helix structure complemented by four nucleic acids—adenine, thymine, guanine, and cytosine, all connected by hydrogen bonds. Adenine and thymine are connected by two hydrogen bonds while guanine and cytosine by three hydrogen bonds. The various sequences of these four nucleic acids carry the information. In developed species, all the genetic information is not contained in one single DNA molecule but is distributed in several containers made of proteins called chromosomes. Chromosomes are encapsulated in the nucleus of every cell. Each species of organism has different number of chromosomes. During cell division, the chromosome is duplicated and shared to the newly formed cell. If the gene fails to make exact copy of the cell such that the cell division is abrupt and uncontrolled, we call it cancer.

Metabolism has a double role, extracting energy from organic materials by oxidization and constructing cells by using the energy. In eukaryotic cells, a membrane-enclosed organelle called mitochondria generates adenosine triphosphate or ATP which is the source of energy. The main set of chemical reactions involved in the production of ATP is known as citric acid cycle. This is done by the proteins present in the inner membrane of mitochondria which oxidize major products of glucose. Mitochondria play a crucial role in the control of cell cycle and cell growth as well as cell death. The growth, aging, and death of any organism are controlled by mitochondria. The number of mitochondria in a cell depends on the species and on the particular tissues. Mitochondria have their own genetic system, i.e., they replicate themselves independently without the help of DNA.

Without going to further details of biochemistry and molecular biology, we understand how complicated process is life, especially multicellular evolved life.

All biochemical reactions take place in a synchronized way, and if one process fails, the entire system gets into problem. The biochemical processes are determined by the number of electrons in carbon that are free to combine with other atoms. This is called valency and the valency of carbon is four which means carbon has four electrons free to combine with other atoms including another carbon atom. Carbon is the fourth most abundant element in the whole universe. Apart from the fact that carbon is the fourth most abundant element in the universe, it has the advantage of chemically combining with several elements to form a large variety of molecules. Any other element cannot make so many large molecules that are needed for an organism. Although silicon is chemically similar to carbon and can form long-chain molecules called polymers, silicon atoms are bigger than carbon atoms and so need more binding energy. Carbon compounds are easily and highly soluble in water. Therefore, carbon-based chemistry is most favored for life, at least on the Earth. However, the chemical bonds between atoms and molecules need a range of pressure and temperature beyond which the molecules break down. For example, at about 160 °C, the molecule adenosine triphosphate or ATP that governs metabolism breaks down. Therefore, theoretically, life cannot exist above a temperature of 160 °C. Of course, many other important molecules may break down or chemical reaction may cease bellow this temperature. It depends on the specific complex molecular structure of a living specimen. A fish cannot survive above a temperature of 40 °C, whereas a microbe such as bacteria can survive at a temperature as high as 75–90 °C. Similarly, metabolic activities may stop at temperature as low as -10 °C. However, there are some forms of life that can adjust and survive under conditions that are extreme for normal life. We shall discuss it at the end of this chapter.

Life Under a Changing Earth

The environment in the Earth was not always the same. It was much different a few billion years ago. The Sun was born approximately 4.6 billion years ago, and the formation of the Earth was complete within a few hundred million years after that, approximately 4,567 million years ago. The oldest material found is zircon crystals which have an estimated age of 4,300 million years. After its birth, the Earth has gone through several changes in its geology that affected the overall climate. The evolution of life is strongly coupled with the geological evolution of the Earth. The geological history of the Earth can be divided into a few primary time intervals known as Eon or Aeon. One Eon lasts for about a billion years. There are four Eons: (1) Hadean, (2) Archean, (3) Proterozoic, and (4) Phanerozoic. Eons are divided into eras and each era has a duration of 100 million years. Eras are divided into periods that last for 10 million years. Periods are divided into epochs and ages. Now we shall briefly discuss the geological and the biological events that occurred during these eons and eras.

Hadean Eon: 4,567–3,800 Million Years Ago

The most primitive meteorites or Chondrites are the oldest rock in the solar system. About 80 % of the mass of Chondritic meteorites is composed of Chondrules. Chondrules are tiny spherical dust particles containing minerals in crystalline form. They have a layer structure. Clumps of stellar dust when heated and get partially melted form the spherical Chondrules. The Chondrules within the Chondritic meteorites carry the clues about the formation of the solar system. Chondrules started forming in the solar nebula about 4,650 million years ago. The solar system was formed about 4,567 million years ago. The Hadean Eon started from this time. It is believed that the Earth was formed about 4,500 million years ago. About 4,450 million years ago, just 50 million years after the formation of the Earth, the Moon was formed out of a collision with another planetary body as big as Mars. This object is sometimes called Orpheus or Theia. At this time the Moon was only 64,000 km away from the Earth, and the Earth was rotating so fast that the duration of a day was just 7 h. Subsequently, the tidal interaction of the Moon stabilized the rotation of the Earth. Much of the light gases such as hydrogen and helium escaped from the Earth during this time. At the beginning, the Sun was only 70 % as bright as it is today. Therefore, the surface temperature of the Earth should have been below the freezing point of water. Then how the liquid water existed in the early Earth? We shall address this issue in detail in the next chapter.

About 4.0–3.9 billion years ago, the cataclysmic meteorite bombardment took place. This cataclysmic event occurred most probably by the debris produced out of a collision of Mars with an object as large as 2,000 km in diameter. By this time the Moon moved to a distance of about 282,000 km from the Earth, and it was already tidally locked so that only one side of it faces the Earth permanently. Both the Earth and the Moon suffered from this meteorite bombardment. After the rate of meteorite shower reduced significantly, the carbonate minerals start forming and the atmosphere of the Earth was dominated by carbon dioxide, methane, ammonia, and water vapor. Obviously, the planet was extremely hostile for life during this period.

Archean Eon: 3,800–2,500 Million Years Ago

About 3.8 billion years ago, the Archean Eon started and the crust of the Earth started solidifying from molten state. About 3 billion years ago, the Earth sufficiently cooled down so that the solid land started forming. The water also started condensing into liquid form. By then, the day has become longer to 15 h and the Sun brightened up to 80 % of its present value. About 3,500–3,000 million years ago unicellular life appeared on the Earth. The atmosphere became rich in nitrogen—about 75 % of the total volume—and the amount of carbon dioxide reduced to 15 % of the total volume of the atmospheric gas. The dipole magnetic field of the Earth also came into existence during this time. In spite of an apparent hostile environment, life was originated on the Earth about 3,800–3,500 million years ago in the form of single-cell prokaryotes. The oldest prokaryotes found in fossils are as old as 3,500 million years. These prokaryotes survived by anoxygenic photosynthesis as there was no free oxygen present in the atmosphere. Purple and green bacteria used the solar energy to survive by such anoxygenic photosynthesis method.

Proterozoic Eon: 2,500–542 Million Years Ago

Proterozoic Eon is divided into three eras: Paleoproterozoic Era, Mesoproterozoic Era, and Neoproterozoic Era.

Paleoproterozoic Era started 2,500 million years ago and lasted for 900 million years. The first 200 million years of Paleoproterozoic Era is called Siderian Period (2,500–2,300 million years ago). About 2,500 million years ago, oxygen-producing prokaryote bacteria—Cyanobacteria—appeared. During this period stable continents appeared for the first time and free oxygen was available in the oceans and in the atmosphere. This oxygen poisoned the anaerobic organisms. During 2,500 million years to about 550 million years ago, Earth's atmosphere underwent a drastic change. It became oxygen rich. Cyanobacteria are the organisms mostly responsible for the existence of chloroplasts within eukaryotic cells. Cyanobacteria use oxygenic photosynthesis to produce oxygen. Some of them are capable of nitrogen fixation by which molecular nitrogen forms ammonia, nitrides, and nitrates—essential for the growth of plants. Nature favored organisms capable of oxygenic photosynthesis. Therefore, by natural selection, the evolution of life on the Earth was driven by oxygenic photosynthesis. Along with the act of Cyanobacteria, volcanic activities separated carbon and oxygen from carbon dioxide and released good amount of oxygen. Today, atmosphere of the Earth contains about 20 % of oxygen which was mainly produced by Cyanobacteria. Also, 20–30 % of today's photosynthetic products on the Earth are due to Cyanobacteria. Cyanobacteria are found in fossils as old as 2,800 million years. During the Rhyacian Period (2,300–2,050 million years ago), organisms with mitochondria capable of aerobic respiration were born. During the Orosirian Period (2,050–1,800 million years ago) oxygen started accumulating in the atmosphere, and the brightness of the Sun increased to 85 % of its present value. Complex unicellular life appeared during the Statherian Period (1,800–1,600 million years ago). During this period, bacteria and archaeans became abundant.

Mesoproterozoic Era began about 1,600 million years ago and ended about 1,000 million years ago. This era is divided into three periods: (1) Calymmian Period (1,600–1,400 million years ago), (2) Ectasian Period (1,400–1,200 million years ago), and (3) Stenian Period (1,200–1,000 million years ago). During the Calymmian Period, oxygen built up above 10 % in the atmosphere and photosynthetic organisms grew rapidly. Eukaryotic cells appeared during this period, about 1,500 million years ago. Most importantly, the atmospheric ozone layer that protects life against the strong ultraviolet rays of the Sun was formed during this period. About 1,100 million years ago the supercontinent Rodinia was formed and subsequently it broke into several pieces. At the young age, the continental surface was much smaller than the oceanic area. Ocean covered as much as 97 % of the total surface area of the Earth. The Moon introduced tide in the ocean which helped atmospheric circulations, increasing the dry land area, and hence played a crucial role for the transition of life from ocean to land.

Neoproterozoic Era had a duration of about 450 million years (1,000–542 million years ago). About a thousand million years ago, multicellular organisms appeared on the Earth. About 950 million years ago, the day of the Earth had increased to 18 h and the Moon moved to a distance of 350,000 km from the Earth. This is called Tonian Period (1,000–850 million years ago). During the Cryogenian Period (850–630 million years ago), the Rodinia supercontinent broke and the Pannotia supercontinent was formed. About 550 million years ago, during the Ediacaran Period (630–542 million years ago), the Pannotia continent broke up and fragmented into Laurasia and Gondwana continents. The duration of day became about 21 h. Jellyfish-like organisms were developed during this period.

Phanerozoic Eon: 542 Million Years Ago to Present

Phanerozoic Eon is divided into three eras: Paleozoic Era, Mesozoic Era, and Cenozoic Era. Each era is divided into several periods and epochs. Much of the evolution of life took place during this last Eon. During the **Paleozoic Era** (542–251 million years ago) multicellular organisms flourished on the Earth. About 510 million years ago, vertebrates appeared in the oceans and green plants and fungi appeared in the land. The brightness of the Sun was only 6 % less than it is today. The amount of carbon dioxide in the atmosphere decreased during this era. About 416 million years ago, ferns and seed-bearing plants appeared and forest was formed for the first time. The day was about 22 h long. About 400–375 million years ago, land animals and vertebrates with legs appeared. The first amphibians and trees also appeared during this era. The atmospheric oxygen level increased to about 16 % by volume. About 300 million years ago, reptiles were born. The oxygen level increased as high as 30 %. The duration of a day became about 22.4 h as the Earth's rotation slowed down further due to the tidal interaction with the Moon which moved away to a distance of about 375,000 km from the Earth. About 275 million years ago, during the Permian Period, the supercontinent Pangea was formed. The oxygen level dropped to 12 % during this period. However, the worst mass extinction occurred during this period. About 90 % of ocean creatures and 70 % of land life were extinct. The land temperature was raised to 60 °C, while the sea temperature was about 40 °C.

During the **Mesozoic Era** (251–65.5 million years ago), the Pangea subcontinent broke up. Reptiles and small dinosaurs appeared in land during this era. About 200 million years ago, mammals were born for the first time. The Jurassic Period

(199–145 million years ago) was dominated by dinosaurs, giant herbivores, and carnivores. The Earth was so warm that there was no polar ice during this period. About 180 million years ago, the African continent was separated from North America, and about 125 million years ago, African continent and Indian subcontinent got separated from Antarctica. About 105 million years ago, South America was separated from Africa and the Atlantic Ocean was formed.

The **Cenozoic Era** started about 65 million years ago. During this era, about 50–45 million years ago, mountains were formed and Australia was separated from Antarctica. The Himalayas mountain range formed during this time when the Indian continent merged with the Asian continent. The Moon moved further away to a distance of 378,000 km from the Earth and the day became 24 h long. Modern mammals such as camels, horses, and rhinos appeared during this time. About 34 million years ago, the Earth cooled down globally and ice in Antarctica formed permanently.

About 4.4 million years ago, during the Pliocene Epoch (5.3–2.58 million years ago) of Neogene Period (23 million years ago to present), an early hominin genus known as Ardipithecus appeared and Australopithecus, a genus of hominids appeared about 3.9 million years ago. About 2.4 million years ago, *Homo habilis*, the earliest ancestors of human evolved from hominids. The Stone Age began about 2 million years ago. From *Homo habilis*, human and Neanderthal evolved about 700,000 years ago. Finally, about 160,000 years ago, the modern *Homo sapiens* appeared on the Earth.

The Evolution of Life

Once the atmosphere of the Earth became oxygen rich, prokaryotic single-cell organisms appeared about 1,500 million years ago during the oxygenation of the Earth's atmosphere. Multicellular life appeared only about 500–750 million years ago. In a changing Earth, individual organisms that adapted the environment the best survived and evolved. This adaptation by adjusting with the environment triggers evolution. All other species got extinct at some stage.

What is evolution? If some organisms of the same specimen of life live in dry land while the other live in water, both would adjust to the different environments. This will lead to change in their living habits which in turn would change the functioning of their organs. The information will carry on by the gene for generations which in a systematic way would cause a change in their organs in order to make them fit for survival under the existing environment. The gene is responsible for replicating an exact copy, and when a cell divides, it produces exact copies. A slight insignificant deviation from an exact copy due to the response to the environment may not cause any effect during a short period, but it would lead to significant change in the species during a long period, e.g., during a few million years. Ultimately, the difference becomes so much that the two types of organisms—one living in water and the other living in land—would become two separate species although they originated from the same species. The fittest would survive and the rest would extinct. Let us make an analogy. Assume that a class has 500 students but there is only one textbook containing 1,000 pages. So the students need to copy the original book for each of them. Suppose there is no photocopier or any other means to copy the book and so each student has to copy it by handwriting. Now the first student copied the entire original book. However, in doing so he or she mistakenly or intentionally altered ten words. Out of about a hundred thousand words, altering just ten words randomly would not make any difference. The first student then lent his or her copy to the second student. Now this second student who has not seen the original book would not be able to find out the change in the words. While copying from the first copy of the book, the second student also altered another ten words. This continued till the last student copied from the copy of the last but one student. Now if the last student compares the copy he received from the previous student with the original book, he or she would find an entirely different book. If the new book is not interesting, it will be thrown out. But if it is better than the original book, it'll be more popular or advanced. Evolution takes place in a similar manner. The gene which is supposed to make exact copy of a cell makes a small change. This change is inherited to the next generation. In a short period, no change in the organism should be visible. But in a million years, the resulting organism would be entirely different than the original one. In this way, several species are originated out of the same organism. Those species that can adapt themselves with the surrounding environment survive and the other slowly extinct. In this way, complex organisms including human are produced. So a systematic error in the replication process by the gene for several generations causes the evolution. However, if the gene replicates in a rapid but nonsystematic way, we call it cancer. The evolution, however, is not deterministic but probabilistic. If life origins in some other planets, it is not necessary that the evolution of life in that planet would lead to a human- or apelike species. The evolution of life in that planet may not even lead to the appearance of large animals. On the other hand, even if large animals appear, it may not survive long and may get extinct within a few million years. Thus, even if a planet harbors life and life evolves into large animals, it may not produce intelligent life like the human on the Earth. Nevertheless, as the environment influences the evolution of life, life too affects the atmosphere of the planet. Cyanobacteria converted the Earth's atmosphere into oxygen rich that favored the appearance of eukaryotes and ultimately supported the evolution of life from microbes to a very large variety of complex and large animals and plant species.

 On the other hand, mass extinctions of life occurred quite a few times in the past. About 650 million years ago, about 70 % of dominant sea plants were extinct due to global glaciations. About 445 million years ago, glaciations of Gondwana continent caused mass extinction of several marine invertebrate species as well as sea plants. Again, about 375 million years ago, a large number of marine species were extinct. The worst mass extinction on the Earth occurred about 250 million years ago when 90 % of ocean creatures and about 70 % of land-born species, both plants and animals, extinct. The last mass extinction occurred during the end of the Cretaceous

Period (145–65 million years ago). About 90 % of marine species and about 80 % of land species, including the dinosaurs, were extinct about 65 million years ago due to the impact of a large meteor. All these mass extinctions took place due to climatic change and catastrophic events. Extinction of species also occurs due to other reasons such as shortage of food or mass killing or epidemic. The processes of evolution and extinction continue even today.

Origin of Terrestrial Life

In the previous section we have discussed that terrestrial life is made of organic materials. Organic material was not present on the Earth during its birth. So the origin of life depends on the origin of organic materials on it. Therefore, the primary concern is whether the organic compounds were synthesized on the Earth or they were transported from other places in the space. Traces of organic compounds are found on meteorites and comets. If they were synthesized on the Earth, then life, most possibly, had originated on the Earth. If, however, life was transported by comets or asteroids, then it must have survived an extremely cold environment as well as extremely strong radiation of the Sun. It also survived the impact of high-energy cosmic particles which could otherwise break the large organic compounds such as protein and trigger mutation in the gene. It has also survived lack of nutrients for a long period as there was no source of energy. Although many microbes survived such extreme condition on the Earth, there is a limit at which such survival may be possible. As we have mentioned in the previous section, at 160 °C, the ATP molecules responsible for metabolism break down. But still the possibility that life originated in outer space and transported to the Earth is not ruled out. The Murchison meteorite that fell near Murchison in Australia in 1969 contains about 90 different amino acids—the building blocks of life.

In the year 1952, Stanley Miller and Harold Urey performed a famous experiment at the University of Chicago, and they reported in 1953 that organic compounds can be formed out of inorganic compounds under strong radiation of light. Miller and Urey took a mixture of water, methane, ammonia, and hydrogen in their experiment. The chemical mixture was sealed inside a few test tubes connected with each other, and the array of these test tubes was connected with a flask containing liquid water. Then they passed electric spark to the system which resembled lightning in the atmosphere. The water was heated into vapor. After sometime, the mixture was cooled down so that water vapor condensed to liquid water. After a day Miller and Urey found that the mixture became pink in color. After 1 week, it was found that 10–15 % of the carbon contained in methane got converted into organic compounds and 2 % of it formed amino acids, the building blocks of life that make protein in living cells. Carbohydrate and lipid were also formed. This experiment thus confirmed that organic compounds including amino acids can be formed out of inorganic compounds under strong radiation. About 4 billion years ago, carbon dioxide, molecular nitrogen gas, hydrogen sulfide, sulfur oxide, and a few other inorganic gases were released by major volcanic eruptions on the Earth, and these gases mixed up in the atmosphere. Miller and Urey did not take all these gases in their experiment. But it was immediately realized that if these gases were included, more diverse organic molecules could have appeared. The early Earth was not well protected from energetic ultraviolet rays of the Sun as there was no ozone layer. Therefore, during the early period, lightning as well as ultraviolet rays from the Sun supplied the energy for the chemical reactions. Now, in a similar way, organic materials might have been synthesized in asteroids, meteorites, and comets as they are exposed to intense solar ultraviolet rays.

How life was formed out of the organic materials is still unknown. Otherwise, life could have been created artificially in laboratory. Synthetic DNA called XNA is recently produced in laboratory, and it was successfully put into a cell striped out of its original DNA. However, this is far from a living cell originated naturally. At present, there are two types of hypothesis on the origin of life on the Earth. Svante Arrhenius in 1908 suggested that spores that had drifted through space from extra-solar system arrived on the Earth and originated life on it. This is known as "Panspermia theory." In the 1950s of the last century, Sir Fred Hoyle proposed a controversial hypothesis that life might have originated in the interstellar cloud and contaminated the Earth while the solar system including the Earth was passing through such interstellar cloud. Although such a hypothesis lacks any evidence, the presence of organic compounds in meteorites and comets implies the possibility of the existence of Extremophiles outside the Earth. On the other hand, the most acceptable theory is that life originated on the Earth as a consequence of a complex series of chemical reactions that took place simultaneously in the atmosphere of the Earth.

An organism that is capable of synthesizing complex organic compounds such as carbohydrates, fats, and proteins from simple inorganic compounds such as carbon dioxide is called autotroph. Plants in land and algae in water are such organisms that use organic compounds as energy source. On the other hand, the organisms that cannot fix carbon but use organic carbon for growth are called heterotrophs. All animals and fungi are heterotrophs. It is not known which one was originated first, autotroph or heterotroph. It is also not known if life originated in a cold environment or in a hot condition. Nevertheless, it is possible that unicellular microbial life could have originated under hostile environment on both the Earth and on the comets, asteroids, etc., but it needs extremely favorable conditions to survive and evolve into multicellular, complex, and large organisms that we are and we see today on the Earth. We shall discuss it in the next chapter. However, before we end this chapter, we briefly discuss life under extreme conditions. This will tell us that in future even if the Earth's atmosphere becomes hostile to life, even if most of the organisms get extinct, and even if the Earth is destroyed by collision, probably the planet or the debris of it will never become absolutely lifeless. This also supports the "Panspermia theory," which suggests that life was originated outside the Earth and arrived here after traveling under extreme conditions in the space.

Life Under Extreme Conditions

The origin and survival of life need certain favorable environmental conditions that enable the biochemical processes such as metabolism, replication, etc. to sustain. What are the limits or the boundaries of various conditions that make an environment just suitable for life to survive? We know that the majority of life on the Earth is in the form of microbes. The organisms that survive under the limiting conditions at which normal evolved life can certainly not survive are called Extremophiles. Extremophiles exist in conditions far different than the conditions optimal for humans or animals and plants. Temperature, water availability, oxygen supply, exposure to ultraviolet radiation, acidity, etc. are all such conditions that may characterize an extreme environment. Extremophiles can be categorized according to the specific extreme condition under which they survive. Some organisms may survive under more than one extreme condition and they are called Polyextremophiles. How to determine an organism alive under such extreme conditions? If an organism is metabolically active in a given environment, we consider that it is alive and is able to survive irrespective of one or more extreme conditions. Extremophiles may be classified depending on the dominant or extreme condition of the environment, and here we shall point out a few of such classes:

1. **Thermophile and Hyperthermophile**: An organism that requires an ambient temperature between 60 and 80 °C to survive is called Thermophile, while the organism that survives even above 80 °C are called Hyperthermophile. An Archaea named Strain 121 is found to live at 121 °C. Pompeii worms are the largest organisms that survive at 80 °C.
2. **Psychrophile or Cryophile**: Organisms that require optimal ambient temperature bellow 15 °C and even a temperature as low as -20 °C are called Psychrophiles. A large area of the Earth is covered by ice. The Antarctica has the most hostile environment for life. But a wide range of microbes inside the ice and animals on the ice survive. The polar bear survives at a temperature as low as -45 °C. Similarly, a large variety of plants and animals survive deep inside the ocean where the environment is much cooler than that required for the survival of human.
3. **Radioresistant**: Organisms that survive strong radiation especially intense ultraviolet radiation from the Sun are called Radioresistant. The common fruit fly survives 60 times more radiation exposure than that is ambient for human. A red bacterium called *Deinococcus radiodurans* can survive 500 times more intense radiation than the amount a human can withstand. In these organisms, the DNA is tightly coiled and can repair any damage quickly.
4. **Lithoautotrophs**: Microbes that can survive by consuming rocks are called Lithoautotrophs. They derive necessary energy from compounds of mineral origin available in the rocks. These are autotrophs and so can produce organic compounds from carbon dioxide.
5. **Oligotrophs**: Organisms capable of growing in low nutrition are called Oligotrophs.

6. **Xerophiles**: Organisms capable of metabolism with low water are called Xerophiles.
7. **Barophiles**: Organisms that survive under extremely high pressure—as high as 380 times the normal atmospheric pressure—are called Barophiles. Most of the creatures under deep sea and oceans are such type of organisms.

Besides those seven classes of Extremophiles, there are Acidophiles that survive under highly acidic environment. Similarly Alkalophiles survive under strong alkaline condition.

Therefore, life may survive under conditions much different and quite extreme than what we find suitable for ourselves. As a consequence, in future even if the Earth is destroyed, some of the microbes may survive in the rocks without any requirement of oxygen or water or solar energy. They will be spread out in the interstellar medium and subsequently may land into some planet. If the physical conditions in that planet are favorable, these microbes may evolve into complex life as prescribed by the "Panspermia theory" and probably into intelligent life who will ask the question "Is anybody out there?"

This is not only amazing but quite a philosophical perception. According to the Hindu epic, the Bhagavad Gita—"For the soul there is never birth nor death. Nor, having once been, does he ever cease to be. He is unborn, eternal, ever-existing, undying and primeval. He is not slain when the body is slain."

Chapter 8
In Search of Another Earth: An Extremely Rare Planet

> *It took less than an hour to make the atoms, a few hundred million years to make the stars and planets, but five billion years to make man.*
> —George Gamow
> (*In* The Creation of the Universe, 1952)

Conditions for the Origin and Evolution of Life

Although nascent or rudimentary life may originate and survive under extreme environment that is much different than the environment we experience in our daily life, complex life like eukaryotic multigenerational, multicellular organisms require numerous favorable conditions to survive and to develop into large animals. We cannot search for rudimentary or even developed life in all places of the vast galaxy. Further, the only way for us to find out life in outer space is by analyzing the light we receive from the celestial objects, and hence, we can only search for signatures of life in outer space. Fortunately, the presence and evolution of life influence the environment in the same extent the environment affects the survival and the evolution of life. Therefore, the first step towards finding out life outside the Earth is to find out another Earth. Another planet which is very similar to the Earth should have almost all the conditions that favor life to originate and to evolve. But what are those conditions? We are so accustomed to live under this planetary environment that we often do not realize them. We are even not aware of many of the conditions that indirectly preserve our life. If some day, the temperature rises high, we feel terribly hot and we realize that a few more degrees increase in the temperature may cause severe health problems and sometimes even death. On the other hand, we need sufficient protection to survive in the extreme cold of Antarctica. But these are direct effect. There are several indirect phenomena that provide protection to life in long term. Unless we find out a planet having all these favorable conditions, we cannot say that the planet is like our Earth, and so we cannot say with certainty that life exists there. It is worth mentioning here that a rocky planet that has appropriate temperature for water to exist in liquid state is called a habitable planet. But it is not necessary that all habitable planets may harbor life. Let us pose this problem in a different way—let us assume that somewhere in our galaxy there exists another Earth wherein intelligent creatures live and they too are curious to know if there is life beyond their own planet. Suppose that they have similar astronomical and biological knowledge that we have but their civilization is much older and hence more advanced than our civilization and so they have far better technological means to search for life outside their planet. How can such intelligent creatures find out our planet, the Earth? How will they realize that the Earth is not only habitable but also harbors life? Of course they cannot investigate each of the 200 billion stars that our galaxy contains because that will take millions of years even if they have highly advanced technology and sophisticated instruments such as ultrafast computers and extremely powerful telescopes. So the extraterrestrial astronomers will use their knowledge on the galaxy, stars and their properties, etc. and adopt a few clever steps to reduce the time and effort. Their ultimate goal is to find out our Earth and to realize that life exists here on the Earth. In other words, their aim would be at the least to understand that the planet Earth is capable of sustaining life. These steps are related to the conditions needed for complex life to survive and to evolve.

The Galactic Habitable Zone

The environment of our galaxy differs from region to region. The central bulge is dense and hot. There is a massive black hole at the center of our galaxy, and it emits strong X-rays and gamma rays. It also makes the region quite turbulent by strong gravitational interaction. This region is rich in heavy elements and so the stars are very massive—two to ten times more massive than our Sun. Massive stars evolve faster than the lighter ones because they burn hydrogen and helium faster than a solar-type normal star does. At the end of their life, these massive stars explode into supernovae. Further, the rate of collisions by asteroids, comets, and other catastrophic events is also very high at this region. So, even if many of the stars in this region should have planets and many of the planets may have appropriate temperature for liquid water to exist, life cannot survive in such a violent region with frequent catastrophic phenomena. Therefore, the extraterrestrial astronomers would exclude this region.

As mentioned previously, all elements except hydrogen and helium are called metals by astronomers, although a chemist will not call them metals. Hydrogen and helium are produced during the Big Bang nucleosynthesis event, and they are present everywhere in the universe. Now, the amount of metals present in a star plays a crucial role, not only on the physical properties of a star but also on its evolution. Metal-rich stars are called Population I stars and metal-poor stars are called Population II stars. Stars without metal belong to Population III. Most of the

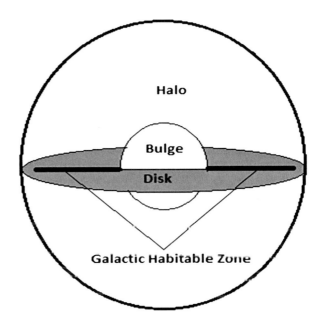

Fig. 8.1 Schematic diagram of the Milky Way Galaxy showing the Galactic Habitable Zone. The disk contains the spiral arms (Illustration by the author)

stars in the galactic halo are Population II stars. They are old and metal poor. As a consequence the halo stars may rarely have planets because it is generally believed that a metal-poor star cannot sustain a proto-planetary disk. Even if such stars have planets, they may not have rocky planets and the amount of carbon present in these planets should not be sufficiently high for the origin of life. Therefore, the extra-terrestrial astronomers would exclude the galactic halo from their search program. Thus, their search will be confined to the galactic disk which contains the spiral arms. Note that our galaxy is a spiral galaxy.

Along the galactic disk, the amount of metals decreases from the inner edge to the outer edge. Therefore, the stars at the outer edge of the disk should not have planets. Even if they have planets, it should be hard to find organic materials essential for life because of the lack of sufficient amount of carbon.

The thick solid lines in Fig. 8.1 thus indicate the region of the galaxy wherein metal-rich stars exist under a relatively quiet environment. This region is called the Galactic Habitable Zone. However, the spiral arms are really not quiet and calm. New stars are continuously formed in these regions, and so the arms are very active in forming stars. Although the spiral arms are not as violent as the galactic bulge, the environment may not favor life to survive for a long duration. Note that it needs sufficiently long time for the evolution of simple unicellular life into complex multicellular organisms. Out of the 200 billion stars in our galaxy, one during 10–100 years explodes as a supernova. This causes intense heat, shock, and radiation in the surrounding region. Therefore, galactic habitability depends on (1) the rate at which stars are formed at a given time in a region, (2) the chance for the survival of the planet or the planetary system itself due to supernova explosions or other catastrophic events, and of course on (3) the amount of heavy elements present in the region where the stars are formed as that determines the formation of planets. The most suitable place is, therefore, the edges and the gaps of the spiral arms of the galaxy. As a consequence, the extraterrestrial astronomers will include our Sun as a suitable target for their next steps because the Sun is situated at the edge of the Orion arm at a distance as far as about 26 thousand light years from the center of the Milky Way Galaxy (see Fig. 1.2). We are located in a relatively quiet and calm region of our galaxy.

Age of the Planet-Hosting Star

Out of a large number of stars at the Galactic Habitable Zone, the extraterrestrial astronomers would pick only those stars that survive for at least four to six billion years as a normal star because the origin of life and especially the evolution of nascent life into multicellular complex life need a few billion years. Now, the age of a star depends on its mass which determines its brightness. The brighter the star, the hotter it is. Hotter stars consume their nuclear fuel faster than the cooler ones do. This is because of the fact that hydrogen fusion process takes different channels of reactions depending on the initial temperature at the core. The initial temperature at the core depends on the mass of the star. The heavier the star, the hotter the core, and so the nuclear reaction in heavier stars takes the channel that makes the consumption of the nuclear fuel faster. Therefore, they exhaust all the fuel in a shorter period, sometimes even in less than one billion years, and start expanding to Red Giant stage. When our Sun will exhaust hydrogen at its core, the core will become heavier and hydrogen will start burning at the shell which would break the hydrostatic equilibrium because outward radiation pressure will be more than the gravitational pull. An expanding Sun will engulf even the planet Mars. However, for our Sun or a solar-type star, it will take another four to five billion years to get all the hydrogen fuel exhausted in the core. The typical age of the Sun and solar-type stars is thus eight to ten billion years. The present Sun is a middle-aged dwarf star. The cooler stars such as the K- or M-type stars live much longer.

So out of all the stars in the Galactic Habitable Zone, the extraterrestrial astronomers would include only those that are not much heavier than our Sun or not much brighter than the Sun so that a long stellar life is ensured. Therefore, their target list would contain stars (1) that are in the Galactic Habitable Zone and (2) that have mass less than or equal to the mass of the Sun. The position of the stars in the Galactic Habitable Zone ensures that they are metal rich. In fact the Sun would get special attention because it rotates in a circular orbit around the galactic center and seldom crosses the active spiral arms of the galaxy.

Now, they would search for planets and planetary systems around the stars that they have listed as their targets. They too may use the same methods for detecting extra-solar planets, e.g., the Radial Velocity method or the Transit method. Since it is assumed that they have more advanced technology than we have at present, they would easily find out the planetary system around the Sun. Further, by using these methods, they would be able to measure the size and mass of all the solar planets accurately. Hence, they would determine that the four inner planets are rocky and the four outer planets are gaseous. They may or may not include Pluto as a planet, but the large distance from the star and the faintness would make Pluto less interesting to them. The next step therefore would be to find out the Habitable Zone or the Circumstellar Habitable Zone of the Sun which is also called as the Goldilocks Zone. They will then investigate the planets that are located within this Circumstellar Habitable Zone.

The Circumstellar Habitable Zone

In Chap. 6, we have already discussed about the Habitable Zone of a star. It is defined as a region at a specific distance from the star such that the temperature is appropriate to keep water in liquid state. A rocky planet within the Habitable Zone is considered as a habitable planet. We know that the Earth is a habitable planet. In fact it is the only habitable planet in the solar system which contains plenty of liquid water, and it is at a distance of one AU from the Sun. We also know that the albedo of the Earth is 0.3 which means 30 % of the total light received from the Sun is reflected back to the outer space by the planet. But the extraterrestrial astronomers would not know that the Earth is habitable. Using the surface temperature of the Sun as 5,505 °C or taking the solar brightness, the extraterrestrial astronomers would find that the Habitable Zone of the Sun extends to a distance from 0.57 to 1.1 AU from the Sun. Therefore, to the extraterrestrial astronomers both Venus and the Earth should appear to be within the Habitable Zone, and Mars should be marginally within it because of its highly elliptical orbit. However, all the planets, satellites, comets, asteroids, etc. have a certain amount of reflectivity. Their surface reflects a certain amount of the sunlight back to the space, and the amount is determined by the albedo of the object. For the planets with atmosphere, the albedo is determined by the properties of the upper atmosphere such as the chemical composition, height, thickness, etc. Therefore, the albedo varies from planet to planet. Now, if the extraterrestrial astronomers consider the albedo of the Earth to be 0.3, then the surface temperature of the Earth at a distance of 1 AU from the Sun would be -18 °C. Clearly at this temperature water would be frozen. So the Earth would not be habitable. Similarly, the albedo of Venus is 0.75. Therefore, at a distance of 0.72 AU from the Sun, Venus would have a surface temperature of -41 °C—much below the water freezing point. Therefore, both the Earth and Venus would appear to be inhabitable to them. Mars is already far away from the Sun and its albedo is 0.2, almost the same to that of the Earth. At a mean distance of 1.52 AU, the surface temperature of Mars should be -59 °C. Therefore, if the extraterrestrial astronomers rely only upon the brightness (or the surface temperature) of the Sun, its distance from the planets, and on the albedo or reflectivity of the planets, they will find no habitable planet around the Sun. What are they missing?

The Greenhouse Effect

The chemical composition and the physical properties of the planetary atmosphere play a crucial role in determining the ambient temperature of the planet. About 30 % of the sunlight gets reflected by the atmosphere of the Earth. The remaining 70 % of the light penetrates the atmosphere and heats up the land or the ocean. The surface then reradiates the heat energy which is now weakened, and so the energy reradiated is in infrared wavelength. When we heat up a metal container by the yellow flame of a gas stove, the container after some time becomes red because the energy reradiated from it has a wavelength associated with red color although the flame emits radiation energy that has a wavelength associated with yellow color. The color changes from yellow to red as the energy weakens. The heat energy transferred by the flame to the container gets partially absorbed by the container and the remaining part is reradiated. Thus, there is a change in energy and consequently a change in the wavelength that determines the color. Similarly, the visible sunlight is converted into invisible infrared radiation due to partial absorption by the Earth's surface. This reemitted infrared radiation gets scattered by the molecules of water, carbon dioxide, nitrogen, etc. in the atmosphere. The photons (particles associated with the radiation) get scattered several times in random directions by these molecules, and ultimately they are redirected towards the Earth's surface instead of going out of it. Thus, water, carbon dioxide, nitrogen oxide, methane, etc. trap the solar energy within the atmosphere and consequently the atmosphere warms up. Water plays the dominant role in trapping the heat with about 60 % contribution while carbon dioxide contributes about 26 % and other molecules such as methane, ozone, nitrogen oxide, etc. also contribute in trapping the reemitted infrared radiation. As a result, the mean surface temperature of the Earth rises by about 33 °C, from -18 to 15 °C, and keeps the water in liquid state. This is known as the Greenhouse effect. For the Earth, the increase in temperature due to the Greenhouse effect varies between 30 °C and 40 °C. For Venus, the Greenhouse effect is much stronger. It increases the atmospheric temperature of Venus from -41 °C to as high as 480 °C, whereas for Mars it raises the mean temperature only by 10 °C. Therefore, if the extraterrestrial astronomers consider only the surface temperature of the Sun and the albedos of the planets, they will not find out any planet in the solar system suitable for habitability. On the other hand, if they take into consideration the amount of various gases present in the atmospheres of the planets and then deduce the albedos as well as the change in the temperature due to the Greenhouse effect, they would find that only the Earth is a habitable planet in the solar system.

Since liquid water needs a solid surface to exist and since Mercury has no atmosphere, the extraterrestrial astronomers would consider only the three planets, Venus, Earth, and Mars, in their investigation. To understand the properties of the atmosphere, they need to take good resolution spectrograph of the infrared radiation of the three planets. Analysis of the infrared spectra would provide the chemical composition, density, thermal structure, etc. of the atmosphere from which they can calculate the amount of heat trapped in the atmosphere and hence the actual environmental temperature of the planet. At present, obtaining sufficiently good resolution spectra of an Earth or a Venus type of planet is not possible by the terrestrial astronomers. This is mainly because at a distance of about 0.5–1 AU, the planets are very faint both in the visible and in the infrared light. Moreover, one has to block the intense star light which is not possible if the planet is too close to the star. For a star fainter than the Sun, the Habitable Zone should be nearer to it, and so the reflected light of the planet cannot be resolved from the star light. As a consequence, it'll not be possible to take good spectra of the planet. However, without sufficient knowledge about the atmosphere, it would not be possible to determine even the freezing and boiling points of water on the surface of a particular planet. For example, the boiling point of water under one terrestrial atmospheric pressure is 100 °C. But if a planet has thicker and larger atmosphere such that the pressure of the atmosphere exerted at the surface of water is double of that on the Earth, then water would boil at 120 °C instead of 100 °C.

The Greenhouse effect also played a crucial role at the time the Earth was quite young. The young Sun was much fainter than it is now. About three to four billion years ago, it was about 30 % fainter than it is today. The brightness of the Sun increases by 10 % during every one billion years. If the albedo were the same, then this would have caused the mean surface temperature of the Earth to reduce by 20 °C from today's value. Therefore, instead of 15 °C, the mean surface temperature of the Earth should have been below the freezing point of water, −5 °C. But there are evidences that liquid water did exist even four billion years ago. So how did the Earth attain the necessary temperature to keep water in liquid state? Although this is a puzzle known as the faint young Sun paradox, it is generally believed that the atmosphere of the early Earth was much different than it is today. The amount of carbon dioxide could have been much more than it is at present. There were methane and ammonia in good amount which played as agents for a much stronger Greenhouse effect that raised the surface temperature above the water freezing point. However, this argument is recently challenged by a team of scientists from CRPS–CNRS University of Lorraine, the University of Manchester, and the Institut de Physique du Globe de Paris. From the ancient fossil soil, they have found that the carbon dioxide level during the early period of the Earth was not sufficiently high to yield the required amount of heat through Greenhouse effect. Methane and ammonia were present during the early period which might have served as Greenhouse agents, but these gases are fragile and easily destroyed by the ultraviolet radiation of the Sun. In the absence of molecular oxygen, there was no ozone layer at the upper atmosphere of the Earth. As a result, the amount of ultraviolet rays impinging on the surface of the Earth was very high which could destroy methane and ammonia in a short period. The next Greenhouse agent that could make the Earth warm is nitrogen. However, this team of scientists analyzed a small sample of air trapped in water bubbles in quartz embedded in extremely old but well-preserved rock in a region at northern Australia and determined the amount of nitrogen dissolved in water. Their analysis indicates that the amount of nitrogen present in the atmosphere of an early Earth was too low to enhance the Greenhouse effect and hence could not warm the Earth sufficiently so that water could exist in liquid state. Therefore, the "faint Sun paradox" remains a puzzle. It may be possible that the planet itself was hotter for a billion of years after its formation. This may be possible if the gravitational potential energy released during the birth was trapped inside the planet and so the cooling rate was slow. After a billion years from now, the Sun will be brighter by 10 % of its present value which will be sufficient to make the ocean water boiling. Therefore, the Earth will become hostile to life much before the Sun engulfs it during the Red Giant phase after four billion years from now.

A solid surface and an ambient temperature suitable for water to exist in liquid form may imply a planet to be a good candidate for harboring life on it. However, it needs many other appropriate conditions for the origin and especially for the survival of complex life. The extraterrestrial astronomers may never realize whether those conditions prevail or not on the Earth. So they would instead look for biosignatures in the spectrum of the Earth. However, we shall discuss some of these crucial conditions which prevail on the Earth and enable life to survive. This will clearly indicate that the Earth is an extremely rare planet in the galaxy because hardly any other planet can have an environment with such a prohibitive combination of so many optimal conditions that protect life.

Water on the Earth: Where Has It Come from?

Ordinary hydrogen atom has one proton and no neutron in its nucleus. However, there are other kinds of hydrogen atoms which have the same number of protons but one or two neutrons in their nucleus. These are called isotopes. Deuterium is an isotope of hydrogen which has one proton and one neutron in its nucleus. Water molecule which is made of two deuterium atoms and one oxygen atom is called heavy water. The ocean water on the Earth contains a specific percentage of heavy water. In fact, any rocky objects, planets, or comets that contain water carry a proportionate amount of heavy water and hence deuterium. Therefore, the amount of deuterium or more specifically the deuterium to hydrogen ratio acts as a tracer for the amount of water present in an object. The ratio of deuterium to normal hydrogen in bulk Earth is found to be about 150 in one million or 0.00015. In the atmosphere of Venus, it is found to be very high, about 0.016, and in the atmosphere of Mars the deuterium to hydrogen ratio is 0.00078, much higher than that on the Earth. This indicates that at one time, Venus and Mars too had plenty of water which was subsequently dissociated and evaporated by the strong ultraviolet radiation of the Sun and escaped from these planets.

Now the Sun and the planets were formed out of the same molecular cloud or solar nebula that consisted mainly of hydrogen and helium. The abundance of helium or any other elements or isotopes is measured with respect to the amount of normal hydrogen present. The solar nebula should have a homogenous distribution of matter with the same elemental and isotopic abundances. But from the amount of helium isotopes that were produced by deuterium burning inside the Sun and also from the measured deuterium to hydrogen ratio in the upper atmosphere of the gaseous planets Jupiter, Saturn, and Uranus, it can be inferred that the deuterium to hydrogen ratio of the solar nebula just at the time the solar system was born was about 25×10^{-6} or 0.000025, about one sixth of that in the Earth. This means the rocky planets Venus, Earth, and Mars are heavily enriched in deuterium. But if the Earth and the Sun were formed out of the same cloud, they should have the same deuterium to hydrogen ratio that determines the amount of water. So the Earth contains water in much greater amount than that was available in the solar system. Where has so much water come from? This is still an unresolved and debatable topic. Seventy percent of the planetary surface at a mean depth of about 4 km is covered by water. Also, the Earth has plenty of liquid water since about 4.4 billion years ago, i.e., since its early age.

A clue to the solution of this enigma appears from the meteorites. A kind of meteorite called carbonaceous meteorite is the oldest object in the solar system. This meteorite carries water in clay minerals and organic materials in macromolecular structure. Both water and the organic materials contain hydrogen isotopes. The presence of organic materials in such meteorites proves that organic compounds necessary for the origin of life could have come from outside the Earth. On the other hand, it shows a deuterium to hydrogen ratio of about 0.00038. The clays, in contrast, show this ratio as almost the same to that present on the Earth. This means the meteorites are rich in deuterium. This enrichment in deuterium in the organic material of meteorites may be due to the chemical process in the interstellar medium prior to the formation of the star and the planetary system. Therefore, water on the Earth has a relationship with the interstellar medium. Comets too carry a high deuterium to hydrogen ratio. However, when a comet approaches the Sun, water vapor sublimates, and thus, comets can contribute only about 10 % of the amount of terrestrial water.

After decades of searching, very recently a group of geologists from the United States has reported the discovery of a vast reservoir of water trapped at a depth between 410 and 660 km beneath the crust of the Earth, at a region of the mantle known as the mantle transition zone. The amount of water found is three times the amount of water present in the oceans. This huge reservoir of water is locked up in a rock known as ringwoodite, a form of olivine under tremendous pressure. This mineral, ringwoodite, was found in an unusual diamond that came out to the surface of the Earth from the mantle through volcanic eruption. Ringwoodite contains about 1.5 % of water. However, the water discovered is not liquid but in a state called hydroxide ions or hydroxyl ions. Hydroxide ion is a negatively charged compound made of one hydrogen atom and one oxygen atom. Two hydroxide molecules combine with one hydrogen molecule to produce two water molecules. This discovery suggests that the Earth's surface water might have come from deep inside it rather than from icy comets or asteroids. Also, the amount of water discovered beneath the Earth may be comparable to the amount of water that was once present in Mars.

Natural Protection of Life on the Earth

Presence of Giant Planets

The specific configuration of the solar system in which the giant planets such as Jupiter, Saturn, and Uranus are in the outer region and the rocky smaller planets are in the inner region plays a crucial role in protecting the Earth as well as life on it. Jupiter has about 0.1 % of the mass of the Sun, and it plays a role of "big brother" for the inner planets. Not only it had helped the formation of the inner rocky planets, but also it captures and even alters the path of comets and asteroids which otherwise would have bombarded the Earth quite frequently. Such frequent collisions with comets and asteroids could have made the Earth's atmosphere hostile for life. An asteroid before hitting the Earth's surface might be broken into dust which would cover the atmosphere. As a consequence, much of the sunlight would be blocked by the thick layer of dust. Such a collision about 60 million years ago, during the Mesozoic Era, is considered to be the reason for the extinction of dinosaur. Collision by a comet may not only destroy life and cause evaporation of the oceans but also may make permanent change in the tilt of the Earth's rotation axis that governs the seasonal changes of the Earth's atmosphere. Because of the presence of Jupiter, the rate of collisions with comets and asteroids originated from the Kuiper Belt and the Oort cloud at the edge of the solar system has reduced significantly and thus protected life on the Earth. In addition to that, Jupiter played an important role in stabilizing the orbit of the Earth which helped in creating a favorable climate for habitability. Many extra-solar planetary systems have been discovered wherein the giant planets are orbiting close to the star. A terrestrial planet in the Habitable Zone of such a system would have suffered frequent collisions by asteroids and comets because in such a system the giant planets would attract them. Therefore, the location of the terrestrial planets and the giant heavy planets in the solar system provides a protective environment for life on the Earth to survive. We are yet to find another planetary system in our galaxy which has a similar configuration. This may be due to our limitation in detecting such planetary system. Most of the planetary systems detected so far have giant planets very close to their parent stars. Certainly such planetary systems are entirely different than the solar system. In future, even if an Earth-size planet in the Habitable Zone of such a system is detected, it would be difficult for life to survive against frequent collisions by comets and asteroids. Therefore, until now, our solar system is unique in its configuration, and the location of the giant planets plays a vital role in protecting life from catastrophic events.

Presence of Terrestrial Magnetic Field

The Sun ejects copious amount of highly energetic charged particles, a significant amount of which is directed towards the Earth. The amount of energetic particles ejected from the Sun varies with the solar activities. These energetic particles could change the chemical condition of the atmosphere and can destroy the organic molecules including the DNA causing mutation and cancer. Therefore, complex life on the surface of the Earth would not have survived for a long period. Then how is life protected? Planet Earth has a dipole magnetic field also known as geomagnetic field. The magnetic South Pole is located near the geographic North Pole, and the magnetic North Pole is located near the geographic South Pole. The axis of this magnetic field (line joining the two poles) is at present tilted by about $11°$ with respect to the rotation axis of the Earth. This tilt changes with time and just like it happens to the magnetic field of the Sun, the magnetic poles of the Earth too flip albeit very slowly. The magnetic field is possibly caused and maintained by the flow of liquid metals in the core of the planet. Its outer shape is controlled by the solar wind that carries the energetic charged particles. This geomagnetic field prevents the energetic charged particles to enter into the Earth's atmosphere. Some of the particles, however, manage to penetrate through the poles causing huge colorful aurora at the South Pole and the North Pole. Thus, the Earth's magnetic field provides a shield to the atmosphere and saves complex life. The presence of a magnetic field, however, is common in solar planets and their satellites. Jupiter's magnetic field is about 20,000 times stronger than that of the Earth. On the other hand, the magnetic field of Mars is extremely weak possibly due to the lack of geological activities. Therefore, Mars is exposed to the hazardous charged particles carried by the solar wind, and that causes the environment of Mars hostile for macroscopic life.

Earthquakes and Volcanoes

Earthquake and volcanic activities play a crucial role in maintaining the Earth's environment favorable for life. In the previous section, we have discussed how carbon dioxide traps the reemitted infrared radiation and warms up the atmosphere. It is a strong agent for the Greenhouse effect. On the other hand, carbon dioxide contributes to the Bond albedo or the reflectivity of the planet. An increase in the amount of carbon dioxide would increase the Bond albedo, and hence, more sunlight would be reflected by the Earth's surface to the outer space. This would cool down the Earth. Therefore, carbon dioxide plays a double role—cools down the atmosphere by reflecting sunlight and warms up the atmosphere by absorbing the reradiated infrared light. Further, photosynthesis process that produces molecular oxygen requires carbon dioxide. Therefore, an appropriate amount of carbon dioxide must be maintained in the atmosphere. However, the atmospheric carbon dioxide mixes up with rain water and forms carbonic acid. Carbonic acid is a weak acid and it dissolves rocks and minerals such as lime stone, quartz, etc. and then gets washed out onto the oceans to get incorporated into marine sediments. In about three hundred to four hundred million year, all the carbon dioxide in the atmosphere and in the ocean would have been removed in such a process. This would have caused major changes in the atmospheric temperature and could have made the Earth hostile for life. Fortunately, this carbon dioxide returns to the atmosphere through plate tectonic that causes earthquake and through volcanoes. Hence, there occurs a cycle of removal and reappearance of carbon dioxide in a time scale of about one million years. This is called the carbon-silicate cycle or simply the carbon cycle. Although early Mars had volcanic activities, the crustal rock of Mars is much thicker than that of the Earth. Therefore, plate tectonic causing Mars quake is not possible. As a result, Mars does not have this important carbon-silicate cycle. Recent research by using the heat distribution at the surface of the Mars indicates that Mars is probably on the verge of plate tectonic. However, plate tectonic in the Earth or earthquake is occurring from the very young age of the Earth. This means, not all the rocky planets can have the balance in the amount of carbon dioxide. The super-Earths which are much heavier than the Earth and should have a thick crust may also not have plate tectonic, and therefore, the carbon-silicate cycle in such planets within the Habitable Zone of their stars should depend only on volcanic activities which may be unlikely because of the depth of the crust. This may be one important concern for life on super-Earths. Nevertheless, without earthquake and volcanoes, the Earth would have become so cool that complex life could not have survived. It is also possible that much of the water on the surface has appeared through volcanoes. Volcanoes also release oxygen by dissociating carbon dioxide. Without earthquakes or volcanoes, the minerals would not have been reprocessed, and thus, the Earth would have been geologically dead like Mars.

Presence of a Large Moon

Among the inner rocky planets of the solar system, only the Earth and Mars have natural satellites. Venus and Mercury have no natural satellite. Incidentally, the Earth has a big and massive Moon that has significant gravitational impact on the Earth. It is quite fascinating to know that even the Moon blesses life on the Earth. The Moon not only causes tides in the oceans which in turn influence the atmospheric circulations, it also had stabilized the rotation of the Earth around its own axis. Before the Moon was formed out of a collision of a Mars size object, the Earth was rotating much faster around its own axis. During the early epoch, the duration of a day was much shorter, about 12–14 h instead of the present duration of 24 h. The Moon slows down the rotation speed of the Earth by transferring rotational energy from the Earth to itself through tidal interaction. This lengthens the duration of day and night and also makes the orbital period of the Moon longer with time. In about one million years, the day or night becomes longer by 16 s due to the effect of the Moon. Also, the tidal force of the Moon on the Earth's oceans was much stronger earlier as the Moon was closer to the Earth. Of course, the geomagnetic field too played an important role in slowing down the Earth's spin rotation.

The Earth's spin axis is at present tilted by an angle of 23.44° with respect to the plane in which it is orbiting the Sun. This tilt causes seasonal variations between the southern and the northern hemispheres of the Earth. If the Moon were not there and if the Earth were rotating faster, then this tilt in the spin axis would have changed in a chaotic way and would have increased significantly within a period of just ten million years. This is because of the fact that in the absence of a massive Moon, the orientation of the tilt or the obliquity of the Earth or any other inner planets would precess rapidly over time due to the gravitational forces of the giant planets such as Jupiter and Saturn. This would have given rise to severe environmental change and might have caused hostile atmosphere for life. The lunar tidal interaction with the Earth prevents the spin axis precession rate to become high. The spin axis of the Earth precesses around a complete circle in about 25,800 years due to the present obliquity of 23.44° with respect to the orbital plane. We know that during December winter appears in the northern hemisphere and summer appears in the southern hemisphere of the Earth. The northern hemisphere experiences summer and the southern hemisphere experiences winter during June. After about 12,000 years, this precession will cause the opposite—the northern hemisphere will experience summer in December and winter in June, while the southern hemisphere will experience summer in June and winter in December. If the Moon were not there, such changes would have happened quite chaotically and during a shorter interval of time due to the chaotic change in the precession of the Earth's axis.

Due to the gravitational tug of war between the tidal interaction of the Moon and the interactions with Jupiter, Saturn, etc., the obliquity or tilt angle of the Earth's spin axis also changes. It moves back and forth from 22.2° to 24.6°—only about 2.4° over a period of about 41,000 years. At present the tilt angle is reducing towards its minimum value. The presence of a massive Moon is responsible for such a small change in the tilt or the obliquity of the Earth's spin axis. If the Moon were not there, this change in the obliquity could have been quite large. The chaotic change in the obliquity or tilt in Mars is already observed. Mars' obliquity or the tilt in its spin axis changes between 15° and 35° within a period of one tenth of a million years and between 0° and 60° over ten million years. This is due to the fact that Mars has two very small satellites which cannot influence the planets dynamics. Also Mars is nearer to Jupiter. The orbit of the Earth too precesses and the eccentricity changes over time. At present the orbit of the Earth around the Sun is almost circular, and therefore, throughout the year the Earth remains almost at the same distance from the Sun. But in a few tens of a thousand years, the near circular orbit will become elliptical. So the shortest and the longest distances of the Earth from the Sun will differ significantly in a year giving rise to extreme climate. Together, the small change in the obliquity of the rotation axis, the orbital precession, and the change in the eccentricity of the orbit explain the periodic occurrence of the ice age in the Earth through the Milankovitch theory. In the absence of a large Moon, all these seasonal changes would have been quite drastic.

Presence of the Ozone Layer

The Sun not only ejects energetic charged particles during its activities which are carried through the solar wind, it also emits high-energy gamma rays, strong X-rays, and intense ultraviolet rays. These rays can photodissociate the molecules and ionize the atoms in the atmosphere which then would escape the Earth making the atmosphere thin over a long period. If gamma ray or strong X-ray passes through a living cell, the water molecules surrounding the DNA get ionized. The ions then react with the DNA molecule and mutate it causing cancer. Intense ultraviolet ray can cause severe sunburning and skin cancer. Therefore, a moderate exposure to these harmful rays would have destroyed complex life. What protects us and the atmosphere? It is the ozone layer in the upper atmosphere that saves us. The ozone layer was detected in 1913. This protective layer exists at a distance of 20–40 km above the Earth's surface. It absorbs the gamma ray and the X-ray at the top of the layer itself, and it absorbs the ultraviolet ray at about 5–10 km deeper into the layer.

In the previous sections, we have seen how several celestial arrangements protect life on the Earth. In addition to that, life itself protects it. Ozone is a gas consisting of three oxygen atoms (O_3). However, as we have discussed in the previous chapter, the early Earth did not have free oxygen molecules in its atmosphere. The oxygen molecules (O_2) are produced through photosynthesis by Cyanobacteria and the plants. This oxygen is essential for metabolism and we all live on oxygen. The oxygen molecules in the upper atmosphere get dissociated into oxygen atoms (O) by the solar ultraviolet ray. One oxygen molecule then combines with one oxygen atom to form one ozone molecule. Thus, the oxidization of the atmosphere by organisms created the protective shield of ozone layer which saves complex eukaryotic life. On the other hand, Mars does not have an ozone layer, and so the ultraviolet rays of the Sun penetrate deep into the atmosphere and photodissociate and ionize the molecules and the atoms. That is why a trail of an ionized gas is observed following the Mars.

Since the ozone layer at the uppermost atmosphere of the Earth is caused by living organisms, ozone acts as a crucial biosignature for the Earth. Therefore, if in future we are able to obtain the spectrum of the reflected light from an Earth-like planet and if the spectrum shows deep absorption by ozone in the infrared, then it would certainly indicate the presence of life, in any form, on that planet. Once we discover the potential rocky planets in the Habitable Zone of preferably a solar-type star, the next attempt therefore must be to find out the signature of ozone. Of course, the presence of ozone in Venus is reported but in tiny amount. This is due to the chemical processes in the upper atmosphere of Venus and not by biological process. On the other hand, methane could provide a possibility of microbial life on Mars. Methane may be produced by chemical process as well, but it gets easily destroyed by the solar ultraviolet rays. The dissociation of organic materials in biological organisms may produce methane that may be detected. However, no Mars probe has detected methane yet.

Extraterrestrial Intelligence

The planet Earth is more interesting and special because not only it hosts life on it but also it is a place where intelligent life, the human beings, exists. We have already mentioned that the evolution is a probabilistic and not a deterministic process. So, if in some planet life originates and evolves, it is not necessary that the evolution will be directed towards creating intelligent life. Further, the surrounding environment may give rise to an entirely different kind of species although the biochemistry would remain the same. This is because of the fact that the evolution is tied up with the response of the organisms to the environment and the environment of other planets, even if favorable for life, may not be exactly the same as it is on the Earth. There is one more possibility. We have learned in the last chapter how organisms such as Cyanobacteria altered the atmosphere of the Earth. A civilized life may alter the environment much more rapidly into a hostile one and may destroy all evolved and complex organisms. Human civilization is just a few thousand years old. But an older civilization may come into a point when it will cause destruction to the whole civilization. From history, we have learned that many civilizations got destroyed even in a few hundreds of years. Therefore, the emergence of intelligent life is a much rarer event than the existence of ordinary, unintelligent life which does not cause harm to the natural environment and to the ecological system. Also, it is a pure chance that evolution would lead to intelligent life.

But what is intelligent life? Again we face a tricky question. For an astronomer who searches for extraterrestrial intelligent life, the answer should be that the species can transmit, receive, analyze, and interpret electromagnetic signals. It must be capable of distinguishing natural signals and artificial signals. According to this definition, intelligent life on the Earth may be considered to exist for only a hundred years. Consequently, only those intelligent habitats who are within a radius of 100 light years from the Sun can detect us right now because electromagnetic signal propagates at the speed of light. With this assumption, astronomers attempt to find out intelligent life in our galaxy. The most famous project that searches for extraterrestrial intelligent life is SETI or "Search for Extraterrestrial Intelligence." The radio waves emitted by TV, mobile phone, military communications, etc. which are all artificial signals should be quite different than the radio signals from celestial objects. However, till date no such signal is detected. On the other hand, radio signal encoded with the position of the Earth is sent by astronomers with the hope that if extraterrestrial intelligent species exist elsewhere and can receive and interpret this signal, they should respond to it. However, the assumptions involved in such methods are debatable and we shall not go into it.

On the other hand, the historians would define intelligent life as old as the civilization. The Babylonian, the Egyptian, the Inca, the Mayan, and the Aryan—all these civilizations were due to intelligent life, but they were not technologically advanced to generate artificial electromagnetic signals, neither had they had the technology to receive signals from extraterrestrial intelligence. The biologists would say that human beings or the *Homo sapiens* were intelligent much before they formed civilizations. In fact the other species Neanderthals were also intelligent but they are completely extinct.

Therefore, two possibilities occur if we have to believe that intelligent life does exist elsewhere beyond our world. Either they are not technologically advanced and so cannot respond to us, or they are so far away—at the other corner of the galaxy that it will take several thousand years before we receive their signal or they receive our signal. A few people advocate a third possibility that the intelligent life may not be interested to communicate with other worlds. Nevertheless, there are quite a few astronomical projects that attempt to detect the presence of extraterrestrial intelligence, and the interested readers can go through the details about the past and present status as well as the history of SETI published in several documents.

Now, if we assume that life originated and evolved into intelligent species in several planets outside our solar system, it is unlikely that their psychological nature would be drastically different than that of us. Therefore, they should also develop curiosity and attempt to detect life outside their own planet. Not only that, an advanced civilization would try to colonize the neighborhood and therefore would extend their activities to other planets. However, we have not yet received any signal that could be interpreted as artificial neither we see any artificial changes in any neighboring region of our galaxy that would indicate activities of extraterrestrial intelligence. "Where is everybody?" Such an argument put forward by the famous physicist Enrico Fermi is known as "Fermi paradox." Unidentified Flying Objects or UFOs are very popular and often generate great interest among common people. But so far reports of all such objects have been found to be a hoax, and we must not be misguided by wild speculations. According to the famous astronomer Carl Sagan, "…we must accept arguments for extraterrestrial visitations to the Earth only when the evidence is compelling." Therefore, we cannot conclude anything regarding the existence of extraterrestrial intelligence at least in our galaxy. As a consequence, the Earth remains the only place known in the galaxy that not only harbors life but also intelligent life.

Epilogue

Why Care for Our Own World

> What is it all but a trouble of ants in the gleam of a million million of suns?
> —Alfred, Lord Tennyson
> (In *Vastness*)

Mediocrity Versus Rarity

Now we know about the Earth, the solar planets, as well as the planets outside the solar system. So we attempt to answer the eternal question "Is anybody out there?" Readers have already noticed that our search for life is confined within our galaxy—the Milky Way galaxy which has about 200 billion stars. There are about 400 billion galaxies in the whole universe. Obviously searching an Earth-like planet outside our own galaxy is not a realistic project because of the vast distance. But our galaxy is a typical galaxy, an ordinary spiral galaxy among the 400 billion galaxies. Therefore, at least our estimation of habitable planets in Milky Way galaxy can be extrapolated for all other galaxies although we cannot confirm it. It should also be emphasized here that we are searching for life similar to that exists on the Earth.

We have now learned that a formidable combination of astronomical and geological conditions has made it possible for the Earth to harbor life on it. This is an improbable coincidence. Can there be any other planets in our galaxy that are also blessed by such favorable coincidence? Of course there are two possibilities—either our galaxy has several planets that harbor life or the Earth is the only planet in the whole galaxy to have life on it. Both possibilities are philosophically interesting. Either life is very common in our galaxy and hence in the universe or we are alone at least in this galaxy. The opinion among astronomers differs. One group of astronomers, the proponents of SETI (Search for Extraterrestrial Intelligence), and those who strongly advocate for the search of intelligent life believe that about 35 million of planets in our galaxy should have intelligent life on it. On the other hand, the other group thinks that the probability of such a rare combination of astronomical and geological conditions is so small that only the Earth harbors life in the whole galaxy. While astronomers Carl Sagan and Frank Drake believed in the "Principle of Mediocrity," Peter Ward and Donald E. Brownlee proposed the "Rare Earth Theory." In 1960, radio astronomer Frank Drake initiated a project called "Project Ozma" which is the first systematic search for artificial signals from worlds beyond our own. Drake used a radio telescope of the National Radio Astronomy Observatory at Green Bank, West Virginia, to scan radio wavelength. In 1961, Drake proposed a formula that attempts to estimate the possible number of planets within our galaxy that may harbor intelligent life with whom communication may be possible. This is known as Drake equation. This estimation by Drake was aimed to garner support for the search of extraterrestrial intelligence, and subsequently project SETI was funded by NASA. According to Drake equation, if N is the number of planets harboring intelligent life that are capable of sending and intercepting artificial radio wave as a tool for communication with other worlds, then

$$N = R_s \times f_p \times n_e \times f_l \times f_i \times f_c \times L,$$

where R_s is the average rate at which stars are formed in our galaxy, f_p is the fraction of the stars in our galaxy that have planets around them, n_e is the average number of habitable planets per planetary system, f_l is the fraction of habitable planets that actually harbor life or support origin and evolution of life, f_i is the fraction of life supporting planets wherein life has evolved into intelligent civilization, f_c is the fraction of intelligent life that has the capacity to communicate, and L is the duration the intelligent life has been sending the communication signal. Note that the quantities denoted by "f" are fractions and so they can have a maximum value 1. Except R_s, f_p and n_e, the other parameters are highly uncertain or unknown. Therefore, we cannot derive with certainty the actual number of planets harboring intelligent life in our galaxy. However, we can, to some extent, estimate the number of habitable planets in the galaxy by reducing Drake equation into the following form:

$$N = n_s \times f_p \times n_e,$$

where N is now the total number of habitable planets and n_s is the number of stars in our galaxy. We know that our galaxy contains about 200 billion stars. About 17 % of them have planets around them. That makes about 34 billion planets in our galaxy. About half of these planets are supposed to be rocky, and so the number of rocky planets in our galaxy is about 17 billion. This estimation comes from the analysis of data obtained by Kepler space telescope. Francois Fressin of Harvard-Smithsonian Center for Astrophysics used the latest data of Kepler to determine that about one in six stars in our galaxy has an Earth or super-Earth. However, the data used by Fressin includes only those planets that are closed to their parent stars. Data from gravitational micro-lens method predicts about 100 billion of Earth-like planets in our galaxy. But these numbers, 17 billion by Kepler or 100 billion by gravitational micro-lens are the number of Earth-size or super-Earth-size rocky planets. Not all of them are habitable planets. Therefore, we may consider that the number of Earth-size or super-Earth-size planets ranges from 17 to 100 billion. But we do not know how many of these planets are habitable. Once the space-bound telescopes such as GAIA obtain sufficient data for the Earth-size or super Earth-size planets in the Habitable Zone, we can have some idea about this. It is worth mentioning here that a large number of planet-hosting stars discovered by Kepler telescope are M-dwarfs. Since M-dwarfs are the faintest stars, the Habitable Zones of this kind of stars are located very close to them. It is argued that planets in the Habitable Zone of M-dwarfs lack hydrogen and other gases since their birth. Even if they contain sufficient amount of hydrogen gas or volatiles, energetic X-rays and ultraviolet rays irradiated over the upper atmosphere could ionize the gas and compel it to escape the planets. Lack of hydrogen will make the planet parched and hostile for life. Further, the M-dwarfs are very active in nature and they often eject hot plasma. As a result, just one strong stellar activity may make the atmosphere very hostile for the survival of life. Thus, M-dwarf stars may not have Habitable Zone at all. As a consequence, the total number of habitable planets will reduce substantially.

Now, habitable does not necessarily mean that the planet can support life as Drake realized it. Therefore, we need to find out the fraction (f_l) of habitable planets that actually support life. This f_l could be a very small number because it is determined by a combination of several astronomical, geological, and chemical conditions. This prompted Peter Ward and Donald E. Brownlee to propose the "Rare Earth Theory." According to this theory, the Earth could be the only planet in the galaxy to harbor developed and intelligent life. They proposed the following modification of Drake equation:

$$N = N_s \times f_g \times f_p \times f_{pm} \times n_e \times f_i \times f_c \times f_l \times f_m \times f_j \times f_{me},$$

where N is now the number of planets in the Milky Way galaxy that harbor complex and evolved life. Here N_s is the total number of stars in our galaxy. This may range from 200 to 500 billion because we are yet to discover all the faint Brown Dwarfs or even the faintest M-dwarf stars that too can have planets. f_g is the fraction of stars in the Galactic Habitable Zone and is about 0.1. f_p is the fraction of stars that have planets and f_{pm} is the fraction of planets that are rocky. n_e is the average number of planets per planetary system that are in the Habitable Zone of the star. f_i is the fraction of habitable planets wherein life originates and exists in unicellular or microbial form. f_c is the fraction of planets where unicellular life has evolved into complex organism, and f_l is the fraction of the total lifetime of a planet during which biological evolution continues. f_m is the fraction of planets that have a large moon, f_j is the fraction of planetary systems that have giant Jupiter type of planets outside the orbit of the planet having complex life, and f_{me} is the fraction of planets that have a large number of species that survive. We have discussed all these essential conditions in the previous chapter. We could have added some more factors such as the fraction of planets that have volcanic activities and earthquake, the fraction of planets that have magnetic field, etc. Ward and Brownlee did not consider the fraction of planets where complex life ultimately evolves into intelligent life. Instead, they considered the total number of planets in the Milky Way galaxy that harbor complex and evolved life. Note that complex life may not necessarily evolve into intelligent life because evolution is not deterministic but probabilistic and it is the fittest, not the intelligent, that survives.

Now from the analysis of Kepler data, we have

$$N_s \times f_g \times f_p = 200 \times 0.1 \times 0.17 = 3.4 \text{ billion}.$$

So there are 3.4 billion planets in the Galactic Habitable Zone. If $f_{pm} = 0.5$ which means half of the planets in all planetary systems are rocky, then we have 1.7 billion rocky planets in the Galactic Habitable Zone. If, however, the prediction of gravitational micro-lens project is correct, then there should be five billion rocky planets in the Galactic Habitable Zone. This is a large number. But not all rocky planets are in the Habitable Zone of the star. In fact the Habitable Zone of a star is very narrow and so n_e should be very small. We have another six uncertain parameters, and each of them is either 1 or less than 1. Even if we assume that all rocky planets in the Habitable Zone are capable of harboring life in simple microbial form, i.e., even if $f_i = 1$, the other five parameters will make the total number of planets having complex life very small. In fact some of the parameters could be so small that in the whole Milky Way galaxy the possible existence of another planet with developed complex life would be zero. Consequently the Earth, our own world, could be the only planet in the whole galaxy to have complex as well as intelligent life.

Without any prejudice to either Drake's argument or Ward and Brownlee's conservative estimation, even if we accept an intermediate value and consider that about one to ten thousand of planets in the galaxy may have complex life on it, the Earth still would be a rare place in the vast Milky Way galaxy. This is because of the fact that even in that situation the average distance between the Earth and the nearest planet that harbors life should be a few hundreds of light years. That makes the Earth a lonely planet if not the only planet to harbor life. Indeed, without going to any mathematical estimation, we can easily realize that the combination of a large number of favorable conditions that has made it possible for life to originate, evolve, and survive on this planet is an extremely rare coincidence.

Epilogue

Potential Threats to Our World

Although several astrophysical and geological conditions favor the survival of life on the Earth, there are many catastrophic events in the space that may destroy life, even the planet. Space could be a dangerous place. There is always a possibility, albeit low, that some natural disaster may destroy the life-supporting environment on the Earth. So even if 10,000 planets in the galaxy have a rare combination of life-supporting conditions, life may not survive for long in many of these planets. Any natural catastrophic event or even a mistake by intelligent life on it may not only destroy life but also damage the whole environment to such an extent that re-origin may not be possible immediately.

Asteroids, comets, and meteors, depending upon their size and composition, may destroy part or even the whole atmosphere. We have already discussed how a large portion of marine and land life including the dinosaurs were exterminated about 60 million years ago due to the collision of a massive comet. Although this comet might not be a large one but was only about 10 km in diameter, it created a crater as large as 180 km in diameter. The energy released due to the impact was equivalent to the explosion of about two million hydrogen bombs. Similar catastrophic events might have caused mass extinction even before. It is speculated that in 1908, a small comet exploded over Tunguska river in Russia destroying about 60–80 million trees over a large area. Although the chance of a collision with such a comet or even with a larger one is extremely low, once in every 40 million years or so, it remains a potential danger for us. Similarly asteroids, even though they are much smaller in size than the comets, can destroy life in a wide spreading region.

Supernova explosion occurs when a massive star became unstable at the end of its life. On the other hand, when two massive stars collide, it gives rise to super-energetic gamma ray burst. Such an event if occurs within a thousand light years from us, it will cause the destruction of the ozone layer, the natural shield that protects the Earth from dangerous ultraviolet rays of the Sun. The afterglow of a nearby gamma ray burst within our galaxy may produce such amount of heat that the oceans may start boiling. On the other hand, the intense gamma rays could destroy the atmosphere permanently. Fortunately, the possibility of such a catastrophic event taking place sufficiently near to us is very small. On the other hand, a normal supernova can adversely affect the atmosphere if it occurs within 30 light years of the Earth. On average, such a possibility arises in every 240 million years. It is estimated that in one billion years, three supernova explosions may occur within 33 light years. But such an event is often unpredictable.

Although our Sun, the parent of the solar system, supports us in many aspects and the only source of energy, it may cause trouble for us. A super giant flare from the Sun may be million times more powerful than the usual flare we are accustomed with. Such giant flares are observed emanating from newly born or active stars. Although there is no reason for our Sun to emit such a destructive flare, there is certainly a chance of a moderately strong flare coming out from it and causing severe damage to the atmosphere. Even a mild solar flare may raise the temperature to such an extent that it can induce massive flood on the Earth. These are some of the threats the Earth may face from the space. However, natural disaster may take place within the Earth, by geological activities. Intense seismic activities or super-Earthquake may give rise to tsunami and may release a huge amount of carbon dioxide and carbon monoxide. A super-Earthquake can change the tilt of the Earth significantly. This may cause severe change in the climate or the seasons. Super-volcanoes such as the one erupted at Lake Toba in Indonesia about 75,000 years ago or the one erupted at Lake Taupo in New Zealand around 26,000 years ago may upset the ecological system of a vast continent.

On the other hand, the chemical content of the Earth's atmosphere has already changed drastically due to industry by human civilization. This has caused global warming leading to melting of polar ice. Although the rate of global warming is slow at present, it may ultimately destroy the carbon-silicate cycle, and consequently the atmosphere may become inhabitable within a few centuries. While the astrophysical catastrophic events and the geological calamity may destroy the whole or at least part of the planet instantaneously, the man-made hazards may lead to a slow but inevitable extinction of life. We have already discussed about the protective shield of the ozone layer extended from 15 to 30 km above the surface of the Earth. Ozone is continuously being produced and destroyed. However, it is a highly reactive molecule. Just one atom of chlorine can destroy a thousand of ozone molecules. This protective layer of ozone is getting destroyed slowly due to the release of man-made pollution containing chlorine and bromine. Deterioration in the ozone layer causes the energetic ultraviolet rays to penetrate deep inside the atmosphere causing damage to the reproductive cycles of plants, algae, etc. and even genetic mutation to multicellular developed organisms. Similarly, deforestation by human civilization due to agricultural activities and by natural factors such as wildfire causes significant changes in the environment. Forests help to perpetuate the water cycle by releasing water vapor back into the atmosphere. Forests also play as a crucial Greenhouse agent by absorbing the reemitted heat from the surface. Photosynthesis by plants contributes to the balance in oxygen molecules in the atmosphere. With the increase in human population, the demand for fossil fuels increases manyfold. When fossil fuels are burnt, sulfur dioxide and nitrogen oxides are released. These gases react with the atmospheric oxygen and water producing sulfuric and nitric acids. Consequently, these weak acids rain down to the Earth and mix up with the ocean or river water and with the soil. Acid rains may cause water toxic to many aquatic organisms and hence affect the other species in the ecological system. Acid rain also damages forest by contaminating the nutrients of plants and trees. On the other hand, the ecosystems of oceans depend on the natural process of organic matter known as phytoplankton. Phytoplankton is broken down by the bacteria that live at the ocean floors. This process produces oxygen out of carbon dioxide and helps the respiration of the ocean bacteria. Too much chemical fertilizations to the crops prevent the bacteria to break down the plankton and slow down the process of conversion of carbon dioxide into oxygen. Therefore, when too much nitrogen is feed to the phytoplankton, oxygen gets depleted in the ocean causing deaths to marine and plant life. So when excessive nitrogen is dumped to oceans, it causes dead zones to the oceans. About 160 such dead zones are found throughout the oceans on the Earth.

Man-made disasters are slow but certain, while natural disasters are instantaneous but comparatively rare. An exponential increase in the human population has caused many species to extinct completely or to become endanger. With the increase in population, the demand and use of fossil fuels have increased in such an extent that during the next few centuries, the planet may be exhausted of all the available fossil fuels. Technological developments have also enhanced the threat to the environment. Just one nuclear war may contaminate the atmosphere to an extent beyond the habitability and genetic mutation afterwards would result into complete extinction of complex life. Even underground nuclear test may cause imbalance in the crust of the Earth causing too many earthquakes that emit a large amount of carbon dioxide. It may also change the tilt of Earth's spin axis. There are many more man made hazards that may cause extinction of life on the Earth within a thousand or million years. If the Earth indeed is the only planet in the galaxy to have complex life, such destructive role by intelligent life itself would make the whole galaxy devoid of life. On the other hand, if there is intelligent life elsewhere, they would be surprised by observing a peculiar atmosphere of the Earth entirely different than the natural atmosphere of any Earth-type planet.

"That Pale Blue Dot": Our Lonely and Only Home

We started this book by addressing the issue on why we should care about other worlds, and we presumed that it was a general curiosity to know if there are worlds beyond our own, how they look like, if there is another Earth in our galaxy, and if life exists elsewhere. However, we have learned an important lesson that our world, mother Earth, is an extremely rare place in our galaxy and possibly the only planet that has the required astrophysical and geological conditions for harboring life. Even if there are other planets similar to the Earth wherein life may exist, the number of such planets in the vast galaxy is extremely small, and so we are alone at least within a radius of few hundreds of light years. There is no other place within the reach of us that can protect us in case the Earth faces any calamity or catastrophic phenomenon that may threaten our survival.

How our home—this great but lonely planet—looks like in the vast empty space? On 13 October 1994, in a public lecture at Cornell University, astronomer and NASA advisor Carl Sagan presented an image of the Earth taken by Voyager I spacecraft when it was about 6 billion km away from us. Voyager I started its journey in 1977. By 1990, it was at a distance of 6 billion km from us, at the edge of the solar system. During 14 February to 6 June 1990, it photographed the solar system and one of the photographs contained a pale blue dot which was immediately identified as our world—the Earth. This tiny dot of light, the only known place that harbors something great, some rare phenomenon, and some unique process—called life—prompted Carl Sagan to realize the inherent philosophy of the existence of an otherwise insignificant celestial object in the vast empty space:

Look again at that dot. That's here. That's home. That's us. On it everyone you love, everyone you know, everyone you ever heard of, every human being who ever was, lived out their lives. The aggregate of our joy and suffering, thousands of confident religions, ideologies, and economic doctrines, every hunter and forager, every hero and coward, every creator and destroyer of civilization, every king and peasant, every young couple in love, every mother and father, hopeful child, inventor and explorer, every teacher of morals, every corrupt politician, every "superstar," every "supreme leader," every saint and sinner in the history of our species lived there—on a mote of dust suspended in a sunbeam.

The Earth is a very small stage in a vast cosmic arena. Think of the endless cruelties visited by the inhabitants of one corner of this pixel on the scarcely distinguishable inhabitants of some other corner, how frequent their misunderstandings, how eager they are to kill one another, how fervent their hatreds. Think of the rivers of blood spilled by all those generals and emperors so that, in glory and triumph, they could become the momentary masters of a fraction of a dot.

Our posturings, our imagined self-importance, the delusion that we have some privileged position in the Universe, are challenged by this point of pale light. Our planet is a lonely speck in the great enveloping cosmic dark. In our obscurity, in all this vastness, there is no hint that help will come from elsewhere to save us from ourselves.

The Earth is the only world known so far to harbor life. There is nowhere else, at least in the near future, to which our species could migrate. Visit, yes. Settle, not yet. Like it or not, for the moment the Earth is where we make our stand.

Very recently, on 19 July 2013, NASA's Cassini orbiter spacecraft that is orbiting the planet Saturn turned back its camera towards us from a position of about 1.2 million km away from Saturn and about 1.44 billion km away from us. The image of Saturn's rings, satellites, and the inner planets was taken in a single frame during a solar eclipse. The spacecraft sent the spectacular image of this great planet, our home which looks like a pale blue dot floating in the vast emptiness of space—a mote of dust so special yet so vulnerable, so crowded yet so lonely, so common yet so unique, so insignificant yet so great—born by a cosmic chance but blessed by many astrophysical and geological happy coincidences, exposed to so many cosmic dangers but survived for so long such that a great phenomenon, a divine process called life originated, evolved and survived to appreciate the beauty of the cosmos and to ask the question "Is anybody out there?"

"The day the Earth smiled." Image of the Earth taken on 19 July 2013 by Cassini spacecraft at a distance of about 1.44 billion km from us (Credit: NASA/JPL-Caltech/SSI)

SO LET US TAKE THE RESPONSIBILITY TO PRESERVE THIS TINY PLACE, THE ONLY HOME WE HAVE EVER KNOWN. LET US CARE FOR OUR OWN WORLD, THE PLANET EARTH AND LIFE ON IT.

Appendix A: Online Resources

1.	www.nasa.gov	Home page of NASA
2.	www.esa.int/ESA	Home page of European Space Agency
3.	www.eso.org/public/	Home page of European Southern Observatory
4.	www.iau.org	Home page of International Astronomical Union
5.	http://kepler.nasa.gov	Homepage of Kepler Mission
6.	http://smsc.cnes.fr/COROT	Homepage of CoRoT Mission
7.	http://hubblesite.org	NASA's webpage for Hubble Space Telescope
8.	http://solarsystem.nasa.gov/index.cfm	NASA's webpage on solar system objects
9.	http://exoplanet.eu	The Extrasolar Planets Encyclopaedia
10.	http://exoplanets.org	List of extra-solar planets
11.	www.seti.org	Home page of Search for Extraterrestrial Intelligence
12.	http://planetquest.jpl.nasa.gov	Planet Quest: The Search for Another Earth, Jet Propulsion Laboratory, Caltech
13.	www.hzgallery.org	On habitable extra-solar planets
14.	www.gps.caltech.edu/~mbrown/sedna; www.gps.caltech.edu/~mbrown/planetlila	Michael Brown's home page on Sedna; Michael Brown's home page on Eris
15.	www.planethunters.org	Interactive Planet Search
16.	https://www.zooniverse.org	
17.	http://universe-review.ca/index.htm	
18.	www.spacescience.org/sitemap	

Appendix B: Some Astronomical and Physical Numbers

Sun's distance from the galactic center	27,384 light years
Sun's velocity around the galactic center	240 km/s
Distance between Earth and Sun (1 AU)	1.4960×10^{13} cm
Distance traveled by light in 1 year (light year)	9.4607×10^{17} cm
Size of the Sun (solar equatorial radius)	6.9599×10^{10} cm
Size of Jupiter (equatorial radius)	7.1490×10^{9} cm
Size of Earth (equatorial radius)	6.3780×10^{8} cm
Mass of the Sun	1.9890×10^{33} g
Mass of Jupiter	1.8986×10^{30} g
Mass of Earth	5.9740×10^{27} g
Mass of Proton	1.6726×10^{-24} g
Mass of hydrogen	1.673532×10^{-24} g
Surface gravity of Earth	980.66 cm/s^2
Luminosity of the Sun	3.839×10^{33} erg/s
Effective temperature of Sun	$5{,}505\ °C$
Equilibrium temperature of Jupiter	$-161\ °C$
Equilibrium temperature of Earth	$-18\ °C$
Mean temperature of Earth	$15\ °C$
Speed of light (c)	$299{,}792{,}458$ km/s
Gravitational constant (G)	6.6726×10^{-8} cm^3/(g s^2)
Stefan–Boltzmann constant	5.67×10^{-5} ergs cm^{-2} s^{-1} K^{-4}
Boltzmann constant	1.38065×10^{-16} erg K^{-1}

Appendix C: Extra-solar Planets with the Calculator

1. Jeans mass:

    Any interstellar cloud having gravitational mass greater than the Jeans mass M_{Jeans} becomes unstable, and external perturbation to the cloud triggers star formation.

    $$M_{Jeans} = \left[\pi kT/(G\mu m_p)\right]^{3/2}/\rho_0^{1/2},$$

    where k is Boltzmann constant, m_p is proton mass, μ is the mean molecular weight, T is the temperature, and ρ_0 is the uniform density of the cloud.

2. Kepler's third law:

    $$P = \left[4\pi^2/G(M+m)\right]^{1/2} d^{3/2},$$

    where P is the orbital period, M and m are the masses of the star and the planet, respectively, G is the universal gravitational constant, and d is the distance between the star and the planet.

3. The variation of the boiling point of water with atmospheric pressure P (in terms of the Earth's atmospheric pressure at sea level) is given by

    $$\text{Boiling point of water} = 373.15 \times \frac{(5.78 - 0.15 \log P)}{(5.78 - 1.15 \log P)} - 273.15°C.$$

4. If T_s is the effective temperature of the star with radius R_s and L is its brightness (luminosity), A is the Bond albedo of the planet, and d is the distance between the star and the planet, then the equilibrium temperature T_p of the planet is

    $$T_p = \left[(1-A)L / (16\pi d^2 \sigma)\right]^{1/4},$$
    $$T_p = T_s(1-A)^{1/4}(R_s/2d)^{1/2},$$

    where σ is Stefan–Boltzmann constant.

5. Greenhouse effect: If T_p is the equilibrium temperature of a planet, T_g is the atmospheric temperature, and τ_g is the infrared optical depth (inverse of the mean free path of photon) from the top to the bottom of the atmosphere, then

$$T^4_g = T^4_p(1 + 0.75\tau_g).$$

Greenhouse temperature is $T_g - T_p$. τ_g for Earth, Venus, and Mars are 0.83, 60.0, and 0.2, respectively.

6. Change of brightness (luminosity) $L(t)$ of Sun at different time:

$$L(t) = L_\odot \left[1 + 0.4\left(1 - t/t_\odot\right)\right]^{-1}.$$

Here, $L_\odot$ and $t_\odot$ are the present luminosity and age of the Sun, respectively. $t = 0$ when the Sun became a star. Therefore, when the Sun became a main sequence star, its luminosity was 30 % less than the present luminosity. After one billion years from now, the brightness will increase by 10 %.

7. Maximum radial velocity V_r of a star of mass M (in solar mass) due to a planet of mass M_p (in Jupiter mass) rotating in a circular orbit with an orbital period P (in year) and viewed inclination angle i

$$V_r = 2840\, P^{-1/3} M_p \sin(i)\, M^{-2/3} \mathrm{cm/s}.$$

Shift in frequency due to Doppler effect caused by radial velocity

$$\Delta \nu = -\nu_{source} V_r \cos\theta/c,$$

where θ is the angle between the direction of the velocity vector and the line of sight and c is the velocity of light.

8. Change in luminosity of a star with radius R and luminosity L due to the transit of a planet of radius R_p

$$\Delta L = L\,(R_p/R)^2.$$

Duration of transit by a planet orbiting with a period P and inclination angle i at a distance d from the Earth

$$\tau = (P/\pi)\left[(R/d)^2 - \cos^2(i)\right]^{1/2}.$$

Minimum orbital inclination angle required for transit to occur

$$i_{min} = \cos^{-1}(R_s/d).$$

9. Radius of Einstein ring caused by gravitational lens:

$$R_E = 8.1(M_L)\,(D_s/8\,\mathrm{kpc})^{1/2}[(1-\beta)\beta]^{1/2}\mathrm{AU},$$

where M_L is the mass of the lens star in solar mass, D_s and D_L are the distance of the source and the lens stars from the Earth in kiloparsec (1 parsec = 3.26 light years), and $\beta = D_L/D_s$.

The micro-lens magnification as a function of time is given by

$$M(t) = \left[u^2(t) + 2\right] / \left[u^4(t) + 4u^2(t)\right]^{1/2},$$

where $u(t)$ is the projected distance between lens and source in units of R_E, Einstein radius.

Duration of the lens event:

$$t = 69.9 M_L^{1/2} (D_s/8 \text{ kpc})^2 [(1-\beta)\beta]^{1/2} / (v/200 \text{ kms}^{-1}) \text{ days},$$

where v is the transverse velocity of the source.

10. Oblateness of gas planets:

$f = (R_e - R_p)/R_e$, where R_e and R_p are the equatorial and polar radii of the planet, respectively.

Chandrasekhar–Darwin–Radau formula:

If the pressure P is a function of density ρ inside the planet and the density distribution can be described by

$P = K\rho^{(1+1/n)}$ where K is a constant and n is called polytropic index that is dependent on the atomic and molecular state of the gas, then

$$f = 2C\Omega^2 R^3{}_e / (3GM_p),$$

where Ω is the angular velocity of the planet around its own axis and C is a constant whose value depends on n. In terms of the moment of inertia I,

$$C = 1.5\left[0.4 + 2.5\left(1 - 3I/2MR^2{}_e\right)^2\right]^{-1}.$$

In terms of the moment of inertia I, Darwin–Radau formula gives

$$f = \frac{\Omega^2 R_e^3}{GM} \left[\frac{5}{2}\left(1 - \frac{3}{2}K\right)^2 + \frac{2}{5}\right]^{-1},$$

where $K = I/(MR_e^2) \leq 2/3$.

Index

A
Acidophiles, 116
Adenine, 106
Adenosine tri-phosphate, 106, 107
Albedo, 32, 37, 39, 41–42, 44, 49, 52, 55, 85, 86, 94, 121–123, 127, 147
Alkophiles, 116
Allotropes, 97
Amalthea, 50
Amino acids, 106, 113
Ammonia, 39, 41, 42, 44, 54, 56, 63, 64, 108, 109, 113, 123
Antimatter, 3
Aphelion, 37
Ariel, 43
Asteroids, 13, 15, 16, 19, 26–28, 35, 37, 39, 46, 47, 49, 50, 54, 56–57, 85, 103, 113, 114, 118, 121, 125, 126, 137

B
Barophiles, 116
Barycenter, 50, 72
Big Bang, 2–5, 10, 63, 104, 118
Black holes, 7, 10, 60, 77, 118
Blue shift, 71, 72

C
Callisto, 40, 50
Carbohydrates, 106, 113, 114
Carbon, 18, 30, 31, 51, 97–98, 100–101, 104–107, 109, 113, 114, 119
Carbon cycle, 128, 138
Carbon dioxide, 31, 33, 34, 36–38, 56–58, 63, 84, 88, 89, 94, 97, 103, 108–110, 113–115, 122, 123, 127–128, 138, 139
Carbon mono-oxide, 37, 48, 58, 63, 97, 138
Cell, 73, 105–106, 109, 111–114, 130
Center of mass, 71–73
Charon, 27, 45, 47, 48, 50, 55
Chondrites, 108
Chondrules, 108
Chromosomes, 106
Chromosphere, 30, 31
CMBR. *See* Cosmic microwave background radiation (CMBR)
Comet Hale–Bopp, 54, 57
Comet Hyakutake, 54, 57
Comets, 2, 13, 15, 16, 19, 22, 24–26, 54–57, 85, 103, 113, 114, 118, 121, 124–126, 137
Comet Shoemaker–Levy, 57
Comet Swift–Tuttle, 57
Corona, 30
Corot, 80, 89, 92, 93
Cosmic microwave background radiation (CMBR), 4
Cyanobacteria, 109, 112, 130, 131
Cytosine, 106

D
Dark energy, 3
Dark matter, 3, 5–7
Deimos, 38, 39, 50
Deoxyribonucleic acid (DNA), 106, 114, 115, 127, 130
Deuterium, 61, 62, 64, 65, 99, 124–125

Diamond, 97–98, 100–101, 125
Dione, 42
Disk
   circumstellar, 10
   proto-planetary, 19, 31, 53, 64, 80, 97, 99, 119
   protostellar, 18–19
DNA. *See* Deoxyribonucleic acid (DNA)
Doppler effect, 71–73, 148
Dysnomia, 49, 50

E
Eclipse, 74, 76, 83–84, 140
EELT. *See* European Extreme Large Telescope (EELT)
Egress, 76–77
Enceladus, 42, 49
Enzymes, 106
Eons, era, epochs, periods, 3–5, 10, 12, 18, 23, 32, 49, 53–56, 68, 73–75, 78–80, 83, 87, 88, 104–105, 107–114, 120, 123, 126–130, 147, 148
Equilibrium temperature, 39, 85, 86, 145, 147, 148
Eukaryotes, 105–106, 109, 112, 117, 130
Europa, 40, 50, 51
European Extreme Large Telescope (EELT), 80
Extinction, 110–114, 126, 132, 137–139
Extremophiles, 114–116

F
Force
   electromagnetic, 3, 4, 61
   gravitation, 2–4, 10, 16–18, 20, 45–47, 50, 53, 55, 129
   strong, 4
   weak, 4, 59

G
GAIA, 81, 89, 135
Galactic halo, 119
Galaxies
   Andromeda, 6–8, 13
   dwarf galaxies, 6, 7, 10
   elliptical, spiral, irregular galaxies, 6–8, 119, 120, 133
   Hubble sequence, 7
   Large Magellanic Cloud, 6–8
   Lenticulars, 7
   Milky Way, 6–9, 11, 13, 29, 119, 120, 133, 136
   satellite galaxies, 6–8
   Seyfert, 7
   Small Magellanic Cloud, 6–8
Gene, 106, 111–113
Geodesic, 3
Grand unification era, 4
Great dark spot, 45
Great white spot, 42
Guanine, 106

H
Hadrons, 4, 5
Halley's comet, 25, 26, 54, 57
Haumea, 28, 47, 49
Helium, 4, 5, 10, 18, 19, 30, 31, 36, 39, 41, 42, 44, 59, 61–63, 88, 108, 118, 124–125
Hellas Planitia, 37–38
Higgs particle, 4
Hormones, 106
Hydra, 6, 48
Hydrogen, 4–6, 10, 12, 18, 19, 30, 31, 37, 39–42, 44, 45, 51, 53, 59–61, 84, 88, 104–106, 108, 113–114, 118, 120, 124, 125, 135, 137, 145
Hydrostatic equilibrium, 16, 28, 53, 64, 120

I
Inclination angle, 73, 74, 76–78, 148
Inflationary era, 4
Ingress, 75–77
Io, 40, 50–52

J
Jean instability, 17
Jeans length, 17
Jeans mass, 17, 147
Jovian planets, 15, 19, 39, 84, 87, 90

K
Kepler space telescope, 89, 134
Kerberos, 48
Kuiper belt, 27–28, 46–49, 52–56, 126

Index

L
Leptons, 5
Lipids, 106, 113
Lithium, 4, 5, 10, 18, 62, 63

M
Magnetosphere, 40, 42, 51
Makemake, 28, 47, 49, 54, 55
Metabolism, 105, 106, 113, 115, 116, 130
Metallic hydrogen, 40
Meteorites, 56, 57, 108, 114, 125
Meteoroids, 37, 56–57
Meteors, 56, 113, 137
Methane, 32, 34, 39, 41, 42, 44, 45, 48, 49, 51, 52, 54–56, 58, 62, 63, 84, 89, 108, 113, 122, 123, 130
Milankovitch theory, 129
Mimas, 42
Miranda, 43
Mitochondria, 106, 109

N
Nereid, 45
Neutrinos, 4–5
Neutron star, 10, 60, 68, 79
Nix, 48
Nucleic acid, 106
Nucleoid, 106
Nucleosynthesis, 4, 5, 10, 63, 118

O
Oberon, 43
Oblateness, 41, 149
Obliquity, 34, 42, 53, 129
Occultation, 74
Olivine, 91–92, 125
Olympus Mon, 38
Orion arm, 7, 13, 120

P
Pandora, 50
Perihelion, 2, 27, 35, 56
Perovskite, 91–92
Phobos, 38, 39, 50
Photosphere, 10, 12, 30, 31, 62–63
Planck era, 4
Planetary migration, 19, 88
Plasma, 5, 30, 40, 135

Plutoids, 28, 56
Polyextremophiles, 115
Population I, II, III stars, 118–119
Prokaryotes, 105, 106, 109, 111
Proteins, 4, 5, 106, 113, 114
Protostar, 18, 62
Psychrophiles, 115
Pulsar, 68, 79, 93, 98

Q
Quarks, 4–5
Quasars, 5, 7, 77

R
Red shift, 71, 72
Red spot, 39
Reference frame, 1–2
Relativity
  general theory, 2–4, 27, 35, 77
  special theory, 1, 2
Replication, 105, 106, 111, 112, 115
Rhea, 42
Ringwoodite, 125
Roasters, 88–89

S
Secondary eclipse, 76, 83–84
Solar nebula, 16, 108, 124, 125
Solar wind, 30, 34–36, 38, 40, 42, 45, 51, 57, 58, 127, 130
Spectrum, 10, 31, 32, 73, 78, 83, 124, 130
Stars
  black dwarfs, 59, 60
  gamma ray burst, 137
  globular clusters, 9, 11
  Herbig stars, 10
  L-dwarfs, 63
  methane dwarfs, 63
  open clusters, 9, 11, 60
  red dwarf star, 60, 62
  red giant, 10, 59, 120
  star-clusters, 7, 9, 11
  supernovae, 9, 18, 30, 118, 119, 137
  T-dwarfs, 63
  T-Tauri stars, 10
  white dwarfs, 10, 59, 60
  Y-dwarfs, 63, 64, 80
Styx, 48
Surface gravity, 32, 35–37, 42, 44, 52, 63, 145

T
Terrestrial planets, 15, 57, 87, 91–93, 105, 119, 123, 126
Tethys, 42
Theia, 108
Thermophiles, 115
Thirty Meter Telescope (TMT), 80
Thymine, 106
Tissue, 105, 106
Titania, 43
Titius–Bode empirical formula, 25, 49
TMT. *See* Thirty Meter Telescope (TMT)
Transit, 27, 64, 70–71, 74–77, 81, 83–85, 87–89, 92, 110, 120, 125, 148
Tycho's Comet, 24, 25

U
Umbriel, 43

V
Valency, 107

W
White dwarfs, 10, 59, 60
Wolf–Rayet star, 18

X
Xena, 28, 49
Xerophiles, 116